Y0-CCX-814

ENGINEER'S PROCUREMENT MANUAL FOR MAJOR PLANT EQUIPMENT

A Guide to Principles and Procedures

- *Planning*   • *Specifications*   • *Bidding*
- *Evaluation*   • *Contract Award*

Robert C. Leeser

For book and bookstore information

http://www.prenhall.com

Prentice Hall PTR
Upper Saddle River, NJ 07458

Library of Congress Cataloging-in-Publication Data

Leeser, Robert C.
    Engineer's procurement manual for major plant equipment : a guide
to principles and procedures for planning, specifications, bidding, evaluation,
contract award / Robert C. Leeser.
        p.   cm.
    ISBN 0–13–294711–0
    1. Industrial procurement—Management.   2. Letting of contracts.
I. Title.
HD39.5.L437   1996
658.7'2—dc20   95–48081
                                                                                                    CIP

Editorial/production supervision: Pine Tree Composition, Inc.
Cover design director: Jerry Votta
Cover design: Karen Salzbach
Manufacturing buyer: Alexis R. Heydt
Acquisitions editor: Bernard Goodwin

 © 1996, by Prentice Hall PTR
Prentice-Hall, Inc.
A Simon & Schuster Company
Upper Saddle River, NJ 07458

The publisher offers discounts on this book when ordered
in bulk quantities. For more information, contact:

Corporate Sales Department
Prentice Hall PTR
One Lake Street
Upper Saddle River, NJ 07458

Phone: 800–382–3419
Fax: 201–236–7141
E-mail: corpsales@prenhall.com

All rights reserved. No part of this book may be
reproduced, in any form or by any means,
without permission in writing from the publisher.

Printed in the United States of America
10   9   8   7   6   5   4   3   2   1

ISBN 0-13-294711-0

Prentice Hall International (UK) Limited, *London*
Prentice Hall of Australia Pty. Limited, *Sydney*
Prentice Hall Canada Inc., *Toronto*
Prentice Hall Hispanoamericana, S.A., *Mexico*
Prentice Hall of India Private Limited, *New Delhi*
Prentice Hall of Japan, Inc., *Tokyo*
Simon & Schuster Asia Pte. Ltd., *Singapore*
Editora Prentice-Hall do Brasil, Ltda., *Rio de Janeiro*

CONTENTS

PART 1
PRINCIPLES OF PROCUREMENT FORMATION

2. OVERVIEW OF PROCUREMENT BY COMPETITIVE BIDDING   2–1

3. THE KEY DOCUMENTS   3–1

PART 2
PROCEDURES FOR PROCUREMENT FORMATION

8. DRAFTING THE REQUIREMENTS FOR BIDDING 8–1

9. THE BIDDING PERIOD 9–1

10. RECEIVING AND OPENING OF BIDS 10–1

PART 3
THE ROLE OF SECURITIES IN THE PROCUREMENT

PART 4
SPECIFICATION WRITING AND PRODUCTION TECHNIQUES

PART 5
DOCUMENTS FOR A SAMPLE PROCUREMENT

FOREWORD

I am honored to write a foreword to this book, the *Engineer's Procurement Manual for Major Plant Equipment,* written by my good friend, Robert Leeser, which I found absorbing in style and content. While many books explain the bidding process of construction equipment, this is the first book ever written that covers the entire scope of competitive bidding procedures for the multidisciplinary procurement of major plant equipment in the fields of mechanical, electrical, mining, and process engineering.

Robert Leeser has done an excellent job of providing a comprehensive and practical guide on equipment procurement procedures for large and complex plant equipment. Everyone involved in the procurement process (suppliers, design engineers, administrative staff, purchasing) should benefit from the information outlined in this text. All facets of design, fabrication, delivery, documentation, and communications are addressed. His in-depth knowledge of the bidding process, management methods, procurement techniques, engineering specifications, supporting documents, and reports provided are true-to-life examples of projects and services. These documents are essential to lead to the award of a contract agreement.

Much attention is devoted to the legal requirements of engineering contracts. This includes the contractual obligations and the invalidation of the contract agreement. Engineers will find this book very helpful in dealing with the legal ramifications in undertaking plant and equipment procurement responsibilities. This well written section will be an invaluable referral that is often not found in other reference materials.

Most engineers, when first assigned to undertake a major plant or equipment procurement, quickly realize that the task requires considerably more knowledge and experience than their professional training provided. Incomplete information, resulting in poorly written technical and nontechnical documents, leaves bidders unsure as to what work they are required to perform. Robert Leeser's discussions should prove to be an extremely useful source of information to engineering students seeking advice concerning plant equipment acquisition. Therefore it is an ideal textbook for the engineering curriculum in project management and construction management courses at both the undergraduate and graduate level. Equipment suppliers and contractors will also benefit, because contractual production commitments are specified and can thus be more easily serviced.

I am happy to see the book adopt the International System of Units (SI) as proper units in the sample specifications and equipment data sheets. This should conform with the European Community 1992 and United States trade and commerce. Thus, it will be extremely helpful to suppliers and contractors when dealing with procurements of plant equipment internationally.

This book builds on the author's more than thirty years of experience in a wide-ranging area of project design, engineering, specification, manpower budgeting, cost forecasting, scheduling, and quality control. After twenty-five years of experience in engineering design, manufacturing, and management prior to becoming an educator, I find the book extremely useful in dealing with preparation of engineering contracts, bidding documents, and product specifications in connection with design and research projects. As you read and apply these guidelines on the equipment procurement procedures, you will find numerous practical and helpful suggestions. Even an engineer most experienced with the procurement formation process will find techniques outlined in this book that can help hone certain skills. I highly recommend this text to you as a most comprehensive book on the procurement formation in the engineering profession.

Phillip C. Quo
Professor, Mechanical Engineering
University of Cincinnati

PREFACE

This book provides valuable guidance for anyone wanting to know about engineering procurement formation techniques for the acquisition of major equipment. The guidelines and procedures advanced in this book are designed to effectively fill an important void in available texts concerned with both the technical and commercial matters needed for setting up and executing contracts for the procurement of larger and more complex equipment.

With the current rapid growth of new and high-technology equipment and services, any person assigned with responsibilities to manage a procurement process will be confronted with intense pressures created by the demands for speed, economy, efficiency, and performance, as well as a standard of know-how and expertise that is now far greater than previously encountered. Document preparation procedures conveying the precise bidding and contract requirements, evaluating an ever-increasing number of technical and contractual variables and exercising critical negotiating skills, have become so sophisticated, methodical, and organized, that the task of decision making for equipment selection is now much more demanding than ever previously experienced. In this regard, the process for acquiring highly capitalized equipment is too important to be covered by a user-type purchase order, or the acceptance of a contractor's standard form of terms and conditions contract.

This book has been in the making for a considerable period of time, since there are no universal standards in relation to both technical and procedural matters on how to perform the work. All too often there is a tendency to bypass vital procedures by taking shortcuts in order "to get the project started" as quickly as possible without recognizing that this can lead to serious contractual consequences such as late delivery and/or substantial cost overruns. This book therefore focuses on the fundamental concepts and essential framework of information, which are presented in a systematic and convenient format that is easily readable and understood. The material is arranged so that each chapter can essentially stand alone, and be capable of immediate application. Flow diagrams and forms and documents commonly used for equipment procurements, together with an example of a major equipment acquisition are included in this book to illustrate how the objectives of the detailed subject matter can be implemented.

This book will not only be interesting and helpful to those responsible for the preparation of procurement documentation and managing a competitive bidding process, but also to others who are directly and indirectly involved with major equipment procurements. This includes plant owners who ultimately pay the bills and hopefully reap the benefits from their investment. Last but not least, success or failure of the procurement will be judged, and remains on record, by the people who operate and maintain the equipment.

A book covering a project such as this is the result of extensive experience in the field of engineering contracts for both buying and supplying costly equipment, as well as researching available printed material on contracts and specifications concerning the procurement of this equipment. It has also benefited from contributions made by a group of dedicated professionals in several disciplines.

Upon initiating this work, I offered a co-authorship to my engineering colleague and one-time mentor, Robert Steiner. He, however, declined since he felt that he did not have the time to adequately support the effort. Nevertheless, he has spent an extraordinary number of hours reviewing drafts of my work, and providing innumerable incisive comments that have helped to maintain this book as an objective treatise.

I am also deeply indebted for the assistance I received from Ron Wilson, who willingly devoted so much of his time for the extensive review of my writing and made valuable comments and important recommendations in the commercial areas of the text. I offer my sincere thanks to James Dixon and Edward Van Name for checking the chapter on bid and contract securities and making many suggestions. The contributions of these people and of other groups, including major equipment suppliers, who have been consulted in the compilation of this work, not only made the text eminently more readable, but also helped to produce a much better book than I could have accomplished alone. Nonetheless, I retain full responsibility for the contents of the book.

Sincere thanks also go to Sue Hart, who carried out the sometimes arduous task of word processing the manuscript, including setting up the format and layout, and successfully coping with my constant alterations, which always received the highest priority.

Most of all, I owe a great debt to my wife Mary, without whose steadfast understanding and support I would not have been able to research, write, and complete this work.

Robert C. Leeser
Sydney

INTRODUCTION

WHAT THIS BOOK IS ABOUT

This book is about the preparation of documentation and competitive bidding procedures for the multidisciplinary procurement of major equipment in the mechanical, electrical, mining, and process engineering fields of industry. Elements that make an item of equipment "major" are difficult to define. In this book, major equipment is associated with a substantial financial commitment as well as being of substantial significance to the purchaser's operations and source of business. To be read in conjunction with "equipment" in this book are plant systems such as a steam and power plant, a mining longwall extraction plant, or a materials handling plant.

This book is designed for international as well as domestic procurement and readership. Accordingly, there may be portions that are not routine practice in specific circumstances, and terminology that is not universally common. However, the book's entire content was developed from much practical experience, derives from recognized authorities, and consequently will be of material assistance to all in the engineering profession. Readers are cautioned of course, to check and make certain that in document preparation, all federal, state, and provincial laws and regulations pertaining to the contract are strictly adhered to. Nevertheless, the objectives for setting up an engineering contract should remain the same. They should be sound, clear, effective, and fair to both the purchaser and the contractor.

In this book, the word *contract* is used exclusively for the purpose of obtaining equipment or plant systems, and use of the word *purchase order* is excluded for such action. A purchase order frequently lists on the reverse page, or has attached to it, a set of provisions forming the terms and conditions of the purchase order. Whereas the acceptance of a purchase order does create a contract enforceable at law, it is considered that it may not always adequately address the contractual conditions needed for large and complex equipment procurements.

This book aims to provide a comprehensive and practical hands-on "how-to-do-it" guide covering relevant techniques and methodology for both the key commercial and technical aspects that are needed for setting-up and executing procurements for large and complex equipment. Each stage of the procurement formation process has distinct operational activities that are discussed in relative chapters. These activities are interrelated and compiled in a step-by-step manner in the same sequence in which the work is usually performed, from planning through to the point where the purchaser and the successful bidder sign the contract agreement.

Whereas each stage of the procurement formation process can be classified as important, particular emphasis has been placed on the following five key factors:

1. In-depth planning for greater assurance that the procurement will be successfully executed from concept through to completion.
2. Understanding and compiling the contractual, financial, and technical matters and correctly locating them in the various bidding and contract documents.
3. Procedures for drafting the engineering specification(s) and other supporting technical documents.
4. Detailing the precise bidding requirements so that bidders can identify and process the information and data called for in order to submit a complete and effective offer. This will permit the purchaser to prepare a comprehensive comparative bid evaluation to select a contractor to perform the work.
5. Guidelines on analyzing bids and preparing a bid evaluation and recommendation report which leads to the signing of a formal contract agreement by the parties.

The material has been divided into five parts. Part 1 explains the principles of the contract formation process. It summarizes the basic but essential legal aspects of engineering contracts, presents an overview of the competitive bidding process, and describes the role key documents play in a procurement. Part 2 describes recommended procedures from planning the procurement through to signing the contract agreement. Part 3 discusses the functions of bid and contract securities. These documents are treated separately in this book since contract securities are not strictly contract documents because a third party (a surety, insurance company, or bank) issues them. Part 4 summarizes recommended specification writing and production techniques for technical and commercial documents. Part 5 presents an example of bidding, bid evaluation, and contract documents for the procurement of high-capacity centrifugal air compressors.

Events that follow after signing the contract agreement—such as the implementation of the contract, and administration and project management—have not been included in this book as these topics are covered in numerous existing publications.

This book is primarily concerned with procurements involving the fabrication and delivery of equipment, and for systems where basic elements have already been designed. Installation and commissioning by the contractor is an optional extra scope of work at the discretion of the purchaser. The reason for this is that installation is often undertaken by the purchaser's personnel or by a construction company. In recent times, many construction companies have become more diversified from their traditional base of civil engineering, buildings and construction, and also undertake the installation of large mechanical, electrical, and instrumentation works on a single source of responsibility basis. Engaging a construction contractor can result in considerable cost savings because of the greater and more constant use of the constructor's highly capitalized plant, as well as the rapid transfer of temporary facilities, supervisors, and labor between projects. The disadvantage to the purchaser is that resources are frequently required for the preparation of additional bidding and contract documents, and the supervision and administration of multiple contracts, for example, one for equipment and another for installation.

Where a procurement calls for the installation and performance testing to be performed "by others," it becomes important to retain the contractor's equipment warranty and performance guarantee responsibilities. This is achieved by including a special services provision in the contract documents requiring the contractor to be present at the site during installation and commissioning activities. As this book is concerned with major and complex equipment, the contract should also include provisions to be noted in the technical specification requiring the contractor to conduct a series of training sessions in terms of manhours of training and of furnishing of training materials for operating and maintenance personnel. Reference must be made that these provisions form part of the contract. Unit rates should be quoted in the bid if additional training is needed.

For the sake of simplicity, no mention is made in this book of the *Engineer* because of the need to further define the person's duties and responsibilities if such a term is used in the contract. The Engineer may either be a permanent employee in the purchaser's organization or one specifically appointed by the purchaser to administer the contract on his behalf. If the Engineer is not in the purchaser's employ, such appointment involves establishing a formal Purchaser-Engineer Agreement before work can commence on the contract. Whilst these matters are of considerable importance in the contract formation process, further discussions fall beyond the scope of this book.

The principal parties to the procurement contract referred to in this book are the:

Purchaser. Any company or individual who assumes the duties and responsibilities of the procurement on behalf of the owner. The purchaser may or may not become the owner of the work.

Bidder. Any contractor who lodges an offer following the purchaser's invitation to bid.

Contractor. Any company or individual who performs the work under the terms and conditions of the contract.

Wherever mention is made of the *engineer* with a small "e," it should be taken that this person is either the engineer in charge, or one of a group of engineers assigned to the procurement by the purchaser.

Most of the larger companies and public agencies have a permanently employed contracts manager, multidiscipline engineering groups, and a procurement department. However, many of the smaller manufacturing and process organizations, as well as some mining companies, do not have such staff to handle procurements for their new projects.

Most of these organizations assign a senior staff person who, with the assistance of a small number of capable professionals having considerable diversified experience in the work to be undertaken, performs the function of project manager for the procurement. Consequently it is not the intent of this book to detail recommended company organization structures nor to explain departmental responsibilities for handling new procurements. However, brief mention is made in Chapter 4, of interdepartmental participation and coordination during the planning stage of a procurement.

WHY THIS BOOK WAS WRITTEN

In today's highly competitive environment where industry is rapidly moving away from a labor-intensive base to high technology automation, the successful execution of any major procurement is something that just doesn't happen by chance. Unfortunately, many of the current practices and procedures leading to the formation of a contract for the supply of major equipment, including the rendering of specified services, have not always been adequately addressed. It is one of the areas that can cause delay in the implementation of the contract or cost overrun, or both. This can generally be attributed to a lack of familiarity with the concepts and techniques, the inability to apply these for the work, or simply to inadequate attention. Poorly written documents, technical and nontechnical, containing incomplete information, inconsistencies, omissions, contradictions, or ambiguous statements have left bidders in a state of confusion and unsure as to what work they are required to perform. Bidders may decline to bid because of lack of clarity in the purchaser's documents covering specific requirements. When this happens, the time, effort, and cost for preparing the documents will have been wasted. This book identifies the main problem areas, and provides information leading to clarification and a better understanding of how to overcome these issues.

This book is intended to help those who have become aware that the effort needed to prepare the bidding and contract documents requires more skill and experience than they initially anticipated. They may have already searched for information dealing with both the technical and commercial (contractual/financial) aspects for the formation of the contract but had difficulties finding suitable material. This is because most publications tend to deal with specifications and contracts for civil building and construction projects. Other books have been written on engineering and construction contracts covering all types of projects in one volume. It must be recognized that significant technical and commercial differences exist between the formation and execution of civil engineering, building and construction works, and the design and supply of major equipment for projects such as steam and power plants, mining complexes, or processing facilities. For civil engineering and construction projects, the owner or a consultant frequently assumes responsibility for the complete design of the work and the contractor constructs the work in total conformity with the construction drawings and specifications. For the major equipment procurement procedures described in this book, the purchaser makes full use of the specialized expertise of the contractor and specifies only the parameters for design, operating, and performance of the equipment to be furnished. Bidders are invited to submit an offering they consider to be the most suitable in meeting the purchaser's requirements. The technical, contractual, and financial features of each offer are evaluated to provide optimum contractor selection.

After bids have been submitted, experience has shown that many engineers are unable for a variety of reasons to properly evaluate the offers and, at times, do not complete the appraisals within the bid validity period. A common practice is to award the contract to the lowest quoted price. This book describes the need for a detailed bid analysis and evaluation to be undertaken for all major equipment in order to show that the lowest submitted price is not necessarily the best or most cost-effective offer received. Apart from favorable price and other commercial considerations, the purchaser must carefully assess and optimize several critical technical and contractual features that will be of benefit over the expected operating life of the equipment. Such features include conservative design margins, the type and quality of materials proposed for long-term reliability, heat and material balances, energy efficiency, as well as operating and maintenance costs. Awarding a contract without having undertaken a detailed bid analysis can lead to disagreements when the parties try to settle differences in their understanding of the scope of work, a delay in the completion of the work, and cost overruns incurred by both parties.

The reader of this book must realize that there are many complexities associated with the bidding process for major equipment especially in the upper price end of the market for the purchaser as well as the bidders. The bidding costs are a heavy expense for companies lodging a bid. Also, the preparation of bidding documents, bid evaluations, and lengthy contract negotiations do not come cheaply for either party. For bidders, knowing that many purchasers are sophisticated buyers in a competitive environment puts extreme pressures on their submittal efforts. This becomes most evident when the bidding documents call for a massive input of complex information and documentation. The problem for bidders is further aggravated by having to wait prolonged periods of time before learning whether or not they have secured a contract. In the interim, they are committed to keeping a core project team on hold in case they win the contract. This complicates the planning of future work.

This book is probably unique in that it devotes considerable attention to the need to standardize certain key technical and procedural matters applying to almost all major equipment procurements, thereby avoiding having to prepare and handle ever-changing contract conditions for different contracts. This is a matter that is becoming increasingly important to bidders and contractors when they are required to make a detailed review of widely varying definitions and interpretations of contract conditions for each contract they bid for and undertake. Such matters include:

- *Delivery.* Who is responsible for packing, shipping, and loading and unloading at the point of shipment and at the destination?
- *Completion.* Of what and when?

- *Acceptance.* Of work, part or whole, and when?
- *Equipment take over from the contractor to the purchaser.* When, where, and what?
- *Equipment warranty.* When does it begin; what is involved; what are the requirements for the rectification of defects?
- *Payments.* What are the procedural requirements for billing, and what are the purchaser's obligations with respect to making payments?

Model forms of general (standard) contract conditions have been published by several professional technical institutions for use in mechanical, electrical, and process (chemical and mineral) projects. However, it is important to distinguish clearly the interface between the requirements stipulated in the technical specification (scope of work to be performed by the contractor), and the contract conditions that establish procedural matters to be observed by both the purchaser and the contractor. Since there seems to be no industry-wide standard for drafting technical specifications for major equipment, there is the frequent tendency for engineers to incorporate commercial requirements in a technical specification. Much confusion can also arise when technical scope of work provisions have been included in the contract conditions document. Apart from the intermix of technical and nontechnical provisions in the two documents, which is often the case, differing requirements can, and have been, specified for the same item(s) of the procurement. This can occur when people in different departments in the purchaser's organization prepare their respective technical and nontechnical documents, and fail to coordinate their efforts. Such a situation poses the question: "Where do the contract conditions end, and where does the technical specification start?" This book attempts to answer this question.

One must also have a clear understanding as to how, and by whom, the various documents will be used after award of a contract, a matter frequently overlooked. A mix of commercial and technical provisions has frequently necessitated splitting bound volumes so that only the applicable parts are distributed to the purchaser's, contractor's and subcontractor's personnel. Later distribution of revised documents can create serious problems. This book looks at these problems and leads the purchaser to recognize how important these matters are to him.

Also to help remedy the problems outlined in the previous two paragraphs, Appendix A is a locator guide containing a listing of key topics that are commonly included in bidding, technical, and commercial documents. It is a tabulation for guidance of those preparing such documents.

As there is a worldwide movement toward metrication, this book has adopted the International System of Units (SI). Many countries are already mandated by law to use this system. These SI units are used in the sample specifications and equipment data sheets. In conformance with the SI, *mass* is used in lieu of *weight,* which should be avoided as "weight" is a force, whereas "mass" is the quantity of material. Full details about the SI units used in this book are given in a publication from the International Bureau of Weights and Measures: *Le Système International d'Unités.* There are some U.S. practices that differ from the English translation version of the publication in the following areas:

- The American spelling of *meter,* and *liter* instead of *metre,* and *litre*
- The period or point is used instead of the comma as the decimal marker

Information on the SI for engineering procurements is detailed in Appendix B, including some commonly used conversion factors between the SI and Imperial units.

The definition of trade terms such as "Ex Works" and "FOB" used for the delivery of equipment for both domestic and export contracts has often been a source of confusion. For domestic procurements, the all familiar term FOB can be defined in numerous ways such as FOB *origin* or source and FOB *destination.* Unless properly negotiated and drafted in the contract, serious problems can arise if the parties fail to reach an agreement on their rights and obligations concerning these matters. Where equipment procurements require international transportation, exporters and importers worldwide are faced with national differences in regulations, procedures, and languages. To this end, the International Chamber of Commerce has published a set of international rules called *Incoterms* to standardize the definition of trade terms covering the obligations and responsibilities of both the purchaser and contractor for any sales transactions. The format of the Incoterms is explained in Appendix C, and these trade terms are used throughout this book for both domestic and overseas contracts.

Because there are differing conventions for abbreviation, the ones utilized in this book are tabulated separately in Appendix D, all of which have some established authority.

TO WHOM THIS BOOK IS ADDRESSED

This book is only concerned with the preparation of bidding and contract documents for the procurement of major and complex equipment. It is most likely that the seemingly formidable task of writing the requirements for bidding, the technical specification, and the commercial documents, is something that every engineer will be involved with at some stage of his or her career. All of the practical skills needed for effective performance of this work is something that cannot be formally taught. Following procedural guidelines such as those presented in this book, plus hands-on experience gained from the implementation of previous procurements, is perhaps the sole way of following the learning curve for successfully performing the work. Many who already have had procurement responsibilities know all too well the consequences they had to face when faults were found in the documents during the administration of the contract.

Sections of this book are designed so they may be used in the upper levels of undergraduate and/or postgraduate courses, and will prove to be an extremely useful source of procurement formation information. However, much more than formal schooling is needed for the preparation of technical and commercial documentation for the procurement of large complex equipment. Not only are sound engineering capabilities, and a knowledge of the contractual legal aspects required, but also financial planning, negotiating, and management skills. For planning and setting up the contract, there is the need to know how to coordinate the various procurement activities with the interacting groups such as cost estimating and cost control, multidiscipline engineering, purchasing (also embracing inspection, and expediting), construction, commissioning, and plant operation. Even an experienced engineer seeking advice and guidance to effectively undertake a major equipment acquisition will find this book to be a valuable source of reference.

This book will also be of interest to equipment suppliers and contractors who commit large sums of their company's resources to prepare and submit their offers. It is hoped that they will appreciate the detailed methodology and techniques that are outlined, especially in the area of recommended bid evaluation procedures. This will lead to a meeting of minds between the purchaser and bidder and the avoidance of misunderstandings and other sources of frustration that so frequently arise during the bidding process.

READING THE BOOK

Greatest benefit will be gained from this book when the entire text is read rather than a particular chapter of immediate interest. This is because of the interrelationship that exists between each stage of the bidding process and hence between each chapter.

In the absence of an international standards system, the terms and definitions used in this book are the ones considered to have the most universal acceptance. Equivalent terms and meanings of terms are shown in the following section of the book—*Equivalents and Meanings of Terms.*

In writing bidding and contract documents, different countries have varying conventions applying in particular to meanings of words and phrases. However, maintaining consistency throughout these documents is important especially with grammar, spelling, and punctuation to avoid confusion, misunderstandings, and uncertainties. Any material which is ambiguous, vague in meaning or intent, can be seized upon in the event of a dispute and used as evidence in any claim for additional payments or damages.

In this book, terms beginning with a capital or upper-case letter are intended to have a special meaning such as for key documents or a particular application. When printed with a lower-case initial letter, they have their normal dictionary meaning. This convention should be adopted when preparing bidding and contract documents.

The use of the pronouns "he" and "his" is widespread in engineering documents, particularly those for procurement. For practical purposes, the male pronoun is used in this book where the preparation of documents, or the supply of goods and services, is undertaken by, or on behalf of, the purchaser's or the bidder's/contractor's organization. The usage of "his and her" in the text relates to a person rather than the organization.

Little annoys a reader more than inconstant spelling. The spelling in this book follows *Webster's New World Dictionary of American English, 1988 Edition* published by Simon and Schuster, Inc. The deletion of hyphens for all the "non" and "pre" words may not appease those who may still retain their "old-fashioned" habits in writing the documents for an engineering contract.

EQUIVALENTS AND MEANINGS OF TERMS

Terms and their definition used in various engineering institution publications differ as well as in documents prepared by purchasers and equipment suppliers. With the increasing complexity of major procurement contracts involving various parties to a contract and the issuance of numerous written commercial and technical documents, it is essential that consistency in terminology be maintained in all contract documents.

EQUIVALENT TERMS

The following is a list of terms commonly used in the commercial and technical sections and their equivalents. The terms in bold print are those defined and generally used in the text while the others are considered equivalents.

Bid
  Tender
  Quotation
  Offer
  Proposal
Bidder
  Tenderer
  Offerer
  Person
  Supplier
  Manufacturer
  Contractor
Bidding
  Tendering
Bidding Documents
  Bid Enquiry Documents
  Documents for Bidding

  Tender Enquiry Documents
  Specification
Bidding Period
  Tender Period
Bid Form
  Tender Form
Commissioning[1]
  Start-up
Contract Agreement
  The Agreement
  Instrument of Agreement
Contract Documents For Bidding
  Contract Documents for Tendering
Content Of The Bid
  Content of Tender
  Offer
  Quotation

[1]See footnote[1] on page xxiv.

Contractor
Vendor
Seller
Supplier
Manufacturer
Equipment
Goods
Plant
Equipment Warranty
Defects Liability
Defects Guarantee
Period of Maintenance
Guarantee Period
Correcting Period
Program [1]
Agenda
Milestone Document
Project
Job
Purchaser [2]
Owner
Employer
Principal
Buyer
Client

Purchaser [3]
Engineer
Project Manager
Superintendent
Contracts Manager
Schedule [1]
Fill-in sheets
List
Site
On-site
Project site
Shop
Workshop
Manufacturer's works
Work
Works

MEANINGS OF TERMS

Since, on a worldwide basis, certain words and phrases have duplicate or multiple meanings, this book, to avoid misunderstandings, will utilize the following:

Commercial. In this book, the necessary legal, financial, and nontechnical conditions needed to support the placing of a binding and contractually sound procurement.

Technical. In this book, all those matters that have a profound effect on the integrity of the equipment from the planning, manufacture, delivery, through to equipment handover.

Schedule. A tabulated form issued with the bidding and/or contract documents and used

---

[1]See reference to these terms in Meanings of Terms.

[2]The party to the contract named in the contract agreement who pays for the work under the contract and who will ultimately own the work.

[3]In this book, the person who has authority to act on behalf of the purchaser, named in the contract agreement, in connection with the contract.

to record information to be supplied by the bidder, or a list of items or information that will be supplied by the purchaser.

Program. A descriptive notice of the intended progress of a series of events that are to occur over and during the life of the contract. It is usually developed in the form of a critical path network, simple or PERT, presenting elapsed periods of time, as well as milestone dates which govern the progress of the contract.

Commissioning. Activity including the functions of start-up, as well as those required to demonstrate that the purchased equipment meets all terms and conditions of the warranty.

Warranty. Word indicating the contractor's pledge, under law, that the equipment is as represented, and will be repaired or replaced at contractor's expense if it does not meet specification. Warranty is used in place of the frequently used word "guarantee."

Signing "under hand". Expression used to signify the execution of those written contracts where only authorized or official signatures are required to make them enforceable at law.

Signing "under-seal". Expression used to signify that a corporate or government seal is required in addition to the requirements of "under hand," or "in writing."

Classification of Major Contracts by Method of Payment for the Work or for Evaluating the Contract Price

Lump-sum (contract). Is where the contractor must execute the work for an all-in price specified in the contract regardless of the number of parts or components and/or subdivisions of the work. A lump-sum price is either "fixed" or subject to "price adjustment," whichever is specified in the contract.

Fixed-price (contract). Is where the contract price is "fixed" for the duration of the contract at the time the contract agreement is signed, and subsequently does not change except as a result of a variation to the contract, the terms of which have been agreed on by both the purchaser and the contractor.

Price-adjustment (contract). Includes a price adjustment provision specified in the contract whereby the contract price is increased or decreased as a result of changes up or down usually in accordance with published indices related to base-line labor and materials costs that the contractor incurs during the course of the contract.

Cost-plus (contract). Is where the contractor is reimbursed for the actual costs he incurs in carrying out the work, plus a specified fee arrangement to cover his overhead costs and profit.

Note that a *firm price* refers to a bid price (not a contract price) which the bidder is prepared to hold and enter a contract on it. A firm price subsequently incorporated into a contract price does not necessarily become a fixed price. It can still be subject to price adjustment if the contract so provides. Being a price offered by the bidder, the term "firm price" is not used in this book.

PART ONE

PRINCIPLES OF PROCUREMENT FORMATION

Chapter 1 provides a basic introduction to the elements of engineering procurement contracts. The legal applications to contract law have been condensed as a more comprehensive treatment of the subject could burden the engineer's interests. Numerous references should be consulted on the law in the field of engineering contracts since laws vary from locale to locale. Chapters 2 and 3 further explain the principles of the contract formation process to provide the engineer with information and guidance to proceed intelligently with the preparation of a major equipment procurement.

1

ELEMENTS OF
PROCUREMENT CONTRACTS

1.0   INTRODUCTION

The focus of this book is on the preparation of documentation for the formation of engineering contracts for the procurement of major equipment. Since nearly all of this equipment is purchased by companies on the basis of a procurement contract arising from invited competitive bids, it is highly likely that the reader will have to write, negotiate, and/or handle certain procurement documents at some point in his or her career. However, the task is becoming more difficult each year with the rapid development of advanced and new technologies, products, and services, as well as the never-ending emergence of new government laws and regulations for the enforcement of engineering contracts.

Those who are assigned the responsibility for preparing contracts for major equipment procurements must have not only a substantive knowledge of the technical matters associated with the equipment being procured but also possess a certain degree of understanding of the legal issues that govern the procurement. Unfortunately, the legal aspects of engineering contracts are something that engineers in general have preferred to avoid. All too often, because of failure to understand the important fundamental features of the legal principles of the contract, engineers have found themselves confronted with serious problems. By the time they seek expert assistance and advice, they may have already prejudiced their positions with a procurement agreement that will be enforced by the courts. To help the engineer survive in a legal environment, it is befitting that this book begins with a general overview of the basic legal principles, applicable to most major equipment procurements, that the purchaser, bidders, and contractors must be able to recognize during the formation and course of the contract.

This chapter is not primarily intended for contract specialists (even though they may find it an interesting reference) but rather it is written with the aim of providing the practicing engineer with useful information and a better understanding, as well as to clear doubtful points that often arise between the purchaser, bidder, and contractor. Most of the concepts dealt with are supplemented in later chapters with discussions on the contract conditions relating to these equipment procurements.

The second part of this chapter discusses the main types of engineering contracts, and outlines their essential differences that the purchaser must be aware of when formulating the procurement documentation.

1.1   FUNCTIONS OF THE LAW OF CONTRACTS

The need for the legal enforcement of promises came into prominence when the practice of commercial transactions made on the basis of mutual agreements or contracts collapsed because one party to the agreement defaulted on the promise to do something for the other party. This led to the development of contract law to be established within the legal system so that contractual problems could be dealt with by the courts.

The scope of contract law is now extremely broad and represents a specialized legal field. It is here that the practicing engineer assigned with procurement responsibilities must become sufficiently knowledgeable about the major ground rules of contract law that relate to his particular procurement, and be able to identify those legal issues needing resolution that are outside of his experience and should be left to an attorney.

Numerous reference books on contract law have been published for engineers. Many are free from legal jargon, and contain much interesting and useful information that can be easily understood. At least one such legal reference book should be in the engineer's library.

While laws relating to contracts may differ in detail between countries, the basic principles will generally apply. It should also be recognized that the law of contract in each country is subject to a constant process of evolution in line with new developments in industry and changing customs and practices, and it is recommended that the engineer should endeavor to keep abreast of such changes.

1.1.1   What Is a Contract?

A *Contract* in its simplest form defines a promise, or group of promises, that the law will enforce. The promise is to do, or refrain from doing, a particular thing. Taken in a commercial context, the contract is a document in which the terms of the promise are recorded. Extending this further, the contract is an agreement enforceable at law between two (or more) parties for one party or both parties to undertake or perform a particular thing. In undertaking or performing that something, both parties accept certain responsibilities, and in return, receive certain benefits.

For engineering equipment procurements, a contract is formed when the purchaser (one party) accepts an *offer*, or a bid submitted by a bidder (the second party). The purchaser's *acceptance* of the offer, or bid, must be unconditional meaning that the acceptance must be unambiguous and free of "ifs," "buts," "on condition that," "provided that," and be in complete accord with the terms of the offer. If the bid is not accepted on an unconditional basis, or the purchaser attempts to make any material changes in the bid, there is no real acceptance, and hence no contract is formed. As the purchaser has in effect reopened negotiations, the bidder is entitled, if he wishes, to present a counter offer for the purchaser to accept. In essence, the bidder is now inviting the purchaser to enter into a contract, which, of course, is unacceptable.

The purchaser's acceptance of the successful bidder's offer is usually sent by mail for the bidder's reply to the purchaser's explicit invitation to enter into a contract. Contrary to the general belief among some engineers, the contract is deemed effective as of the date when the letter of acceptance leaves the purchaser's possession by the act of mailing, and not when both parties have executed the formal agreement.

1.1.2   Basic Legal Formation Principles

In order to form a contract that is legally binding, and one that the law will enforce, the agreement must be entered into by competent parties who express total compliance in the form required by law. The law will provide a remedy to the injured party should there be a breach, as discussed in Paragraph 1.1.4.

The following are the main prerequisites to the formation of a binding contract:

- Intent of the contracting parties
- Legal capacity of the contracting parties
- Legality of the agreement
- Valuable consideration given by each contracting party

Intent of the Contracting Parties

An agreement between the two parties does not itself constitute a contract. There must be a definite promisor and a definite promisee, each of whom is legally capable of performing the intended part of the agreement. It is necessary for the two parties to have their agreement be legally binding, that is, that the agreement be written and enforceable at law. There is no contract when a "handshake" agreement is made between two parties who have no intention of making their arrangement legally enforceable for compensation should one party default or fail to fulfill the terms of the agreement. This then becomes an "honor-only" arrangement between the two parties.

Legal Capacity of the Contracting Parties

This matter rarely applies to engineering contracts as it concerns persons who by law are deemed to be incapable of entering into a legal binding contract, namely persons classified as infants, lunatics, those under the influence of alcohol or drugs, and persons under conviction and held in legal custody.

It is mandatory that the person signing an agreement on behalf of an organization is in fact the person authorized to do so. Certain countries have statutory requirements for the name and position held by the person signing on behalf of a party to be noted under the signature, a statement that this person has the authority to sign, as well as an authorized person to witnesses the signature.

Genuine Consent of the Contracting Parties

An essential element in the formation of a valid contract is that the consent of the parties to an agreement must be genuine, voluntary, and without undue influence or animosity. However, a contract can be voided if:

- Fraud occurs when one party willfully makes a statement knowing it to be false or intended to mislead the other party
- Where one party can prove that he or she was forced to enter into the contract under threats of violence by the other party
- A genuine mistake of fact has been made, for example, when a party to a contract was not aware that the other party had been taken over by a company that is not acceptable to the first party for a certain reason (competing line of business).

The courts will not provide relief to a party suffering from its own improvidence, or against error of judgment. For example, if a purchaser procures a component without a statement or warranty of quality from the vendor and subsequently finds the component to be of inferior quality, the purchaser is required to stand by the bargain struck with the vendor.

Legality of the Agreement

Certain contracts may be void by law because of illegal actions, such as an agreement being formed where its purpose is to accomplish something that is contrary to law. Contracts of this kind rarely arise, or should arise, in engineering contracts. A typical list includes an agreement to commit a crime, an agreement to defraud individuals, an agreement to obstruct the administration of justice, and other agreements whose purpose is to bypass the law.

Valuable Consideration

Even though the parties have fulfilled all the above discussed prerequisites, the agreement will not be considered a legal binding contract unless the promise given by each party is supported by a consideration. The consideration that the law has imposed, must pass both ways between two parties. It must be something of value to be given by one party, the promisor, in exchange for the other party, the promisee, to do a particular thing. When applied to engineering contracts, a consideration is something that has been successfully negotiated between two parties in which party A (in this case, the purchaser) gives something to party B (the contractor) who in return performs the matter that was agreed to in negotiation.

In an engineering contract, this consideration is in the form of money paid by the purchaser to the contractor, who in return supplies the equipment and/or performs services.

The consideration must be lawful and specific. A promise with a vague or unsubstantial consideration will not be enforceable. Provided the consideration is clearly specified in the contract, this is what the parties obtain, as the law is only concerned that a consideration must be of some value regardless of magnitude, adequacy, or fairness. The courts will not look into the question of whether such consideration is adequate to match the obligations of one party to the other.

1.1.3   Oral and Written Contracts

Incredible as it may appear, an oral contract can be just as valid and binding as a written contract. A contract made orally is one made by mouth. This is a more accurate expression than the usual term—"verbally." Verbally means by the use of words and orally means by spoken words. Here, the familiar "we are old friends and have worked together for years—so why the need and expense to put it in writing?" can become a most expensive belief in confidence. Should a disagreement eventuate at some later stage, it can lead to a long and extremely costly hearing in the court.

Obviously, there are practical difficulties in enforcing oral contracts especially when a judge in court has no written record of conversations before him to enable him to determine who is telling the truth. Because of this, many states have barred oral contracts.

Many states have statutory requirements that contracts be in writing in order to be enforceable, including the following agreements:

- By exchange of letters that may incorporate some standard form document
- Partly by some written documents and partly orally, as where a bid is submitted and is accepted by facsimile or telephone
- Partly by some document and partly by conduct, as where a bid is submitted and the goods are delivered and accepted
- By a formal written contract

This implies that contracts for major equipment procurements must be made in writing.

1.1.4   Breach of Contract and Remedies

A breach of contract occurs when one party to the agreement fails, without justification or adequate excuse, to perform in accordance with the terms and conditions set out in the contract. Apart from terminating the contract, the party responsible for the breach must answer in some appropriate manner for the injury it created. For major equipment procurements dealt with in this book and discussed in Chapter 5, provisions in the contract provide for the precise consequences to be followed in the event of a breach.

Apart from the provisions referenced in the contract conditions mentioned previously, and also performance bonds discussed in Chapter 13, a breach will normally present the injured party with a number of alternatives to serve as remedies. In practice, however, it is invariably better for the parties to come to an agreement outside the courts rather than involve themselves in costly litigation that can also seriously delay the completion of the contract. Legal sanctions attached to contracts should be used as the last resort only, and then with greatest reluctance. When a mutual agreement on remedial action cannot be achieved, the major course is to sue for damages. To process a claim for damage, it becomes necessary to first establish the grounds upon which the claim is made, and then assess a sum fixed by a court in compensation for such damage or for any bona fide loss.

Damages can be divided into the two categories.

1. *Actual damages.* Designed to recompense the party injured as a result of a breach of contract, with a sum of money for what should have been received, or expected to have been received had the contract been performed in accordance with its terms and conditions.
2. *Consequential damages.* Designed to recompense the injured party with the recovery of a sum of money for all reasonably foreseeable losses as a result of the breach of contract.

1.1.5   Modification of Contract

In order to modify an existing agreement, the modification is only valid and binding if it has mutual assent by the parties. Most written contracts for major equipment procurements contain a provision in the contract conditions (see Chapter 5, page 5–23) detailing procedures on how an amendment is to be made. Any modification must have the consent of both the purchaser and contractor, be set out in writing, and the agreement evidenced by the signature of an authorized person from each party.

1.1.6   Termination of Agreements

A contract may be brought to an end in any of the following ways:

- Through fulfilment by performance
- Through mutual agreement
- Through frustration

- Through operation of the law
- Through breach

Termination through Performance

In order to bring about the discharge of the entire contract, both parties to a contract must fulfill all their obligations, or performance must be complete by both sides, in strict accordance with the terms and conditions of the agreement. Full and satisfactory performance by both parties is the most common way of discharging a contract.

Termination by Mutual Agreement

A contract is an agreement between parties. The agreement may be terminated either wholly or in part by a further mutual consent or agreement between the parties to do so. However, there may be laws that define a mutual consent to terminate an agreement. One of the main reasons for agreeing to terminate the contract is the conduct of one or both parties that adversely affects their contractual relationship.

Termination through Frustration

The term *frustration* is frequently used where a specified condition essential to the performance of the contract (or a situation arises, which without a default by either party, or beyond the control of either party) causes the entire agreement to be dissolved. Frustration can occur during the course of the contract when a supervening event or factor not contemplated by either party causes the required performance to become radically different from the intention of the contract when it was formed. When the required performance can no longer be achieved, thereby making it impossible to complete the contract, the contract is deemed to be terminated. Frustration is often referred to as *force majeure* in engineering contracts stipulating a procedure for terminating the contract in the event of a prolonged force majeure (see Chapter 5, page 5–20).

Termination through Operation of the Law

A contract may be terminated by operation of the law independently of the wishes of either or both parties in the agreement. For example, the contract becomes voidable (or nullified) through factors such as fraud, a legally underaged person being party to the contract, or one of the parties being bankrupt.

Termination through Breach

The substance of a breach of a contract that would cause the injured or innocent party to terminate the relationship must be such that it will defeat or render the object of the contract unattainable. A breach or failure by the offending party would force the injured or innocent party to accept something different from what was envisaged by the contract. Apart from a promisor's (the contractor's) failure to perform in accordance with the terms and conditions set out in the contract, a contract may be terminated when:

- A party to the contract relinquishes its duties under the agreement
- A party to the contract finds itself unable to fulfill the duties specified in the contract
- A party acts in a manner of hindering the other party's performance

The law then gives the injured or innocent party the right to cease performance, and bring the contract to an end. It also allows the injured or innocent party to obtain a judgment for damages arising from the breach.

Every breach of contract will entitle the injured or innocent party to sue for damages, but every such breach will not give cause to terminate the contract. A breach of contract considered to be casual or insignificant, or one being regarded as of insufficient importance in terms of the contract as a whole (for example, a default solely in one section of the contract), will not be a justification for terminating the contract. In such cases, the offending party may be called upon to make a payment to compensate for the disadvantage the breach created in the performance of the contract.

1.2   TYPES OF CONTRACTS

Contracts are generally classed in accordance with the type of reimbursement and purchaser-contractor responsibilities.

The more important types of contract commonly employed within major equipment procurements are:

- Lump-sum
- Cost-plus
- A combination of both

Each type of contract contains defined features covering such items as processing claims for payment to the contractor, contract obligations, contract control procedures, and other elements that may affect the time duration to complete the contract.

A brief description of each of the contracts is given below.

1.2.1   Lump-Sum Contract

The *Lump-Sum Contract* is one in which the contractor agrees to perform a specified scope of work under the terms of the contract for a stated sum of money. Depending on these terms, the lump-sum price is either:

- Fixed, *or*
- Subject to price adjustment

A *Fixed-Price Contract* is where the contract price is fixed and agreed between the purchaser and bidder before the contract agreement is signed no matter what predicaments or expenses are encountered by the contractor for the duration of the contract.

Variations to the contract agreed on by both the purchaser and the contractor are a separate matter and involve the issuance of a change or variation order as discussed in Chapter 5.

The selection of a fixed price contract usually depends on:

1. The amount of costs the contractor is willing to accept as risks for any possible increase in costs such as labor, material, tax, or currency exchange rates that may occur during the course of the contract. The contractors risk also covers possible omissions and misinterpretations of the technical specifications that may result in underestimating costs in the bid price.

2. The time duration for completing the contract usually depends on the country of origin for the supply of materials, labor over this period, and the annual rate of inflation. A contractor would not be expected to accept a long-term fixed price contract when there are probable increases in cost for material and labor.

Should the purchaser insist on a fixed-price contract, the bidder will probably increase the bid price to cover the above contingencies. However, an excessive allowance may make the bid uncompetitive.

A *Contract Subject to Price Adjustment* is a type of contract that allows the contractor to recoup cost variations above those contained in the contract, by reason of any rise in cost for materials, labor, freight and/or other charges during the course of the contract. Methods for calculating these variations must be defined in the contract conditions. They are normally based on pre-agreed formulae utilizing selected published statistics released on a regular monthly or quarterly basis.

1.2.2   Cost-Plus Contract

The term *Cost-Plus* simply means that the contractor is reimbursed for all costs applicable to the work plus an amount or a percentage of these costs for his overhead and profit. The main types of cost-plus contracts are:

- Time and materials
- Cost-plus with percentage fee
- Cost-plus with fixed fee

Unlike a lump-sum contract, no contract price is quoted in a cost-plus contract and there are no contingencies to be covered by the contractor. The contract is usually awarded on the basis of a contractor's submitted schedule of cost elements to complete the contract. To prevent an over-run of costs due to the contractor's inability to perform the work within the submitted cost estimate, the purchaser may impose a predetermined ceiling or price limit on the work and all charges above this price are to be borne by the contractor to complete the contract.

For a *Time and Materials* type contract, time of the contractor's labor for engineering services is paid for at a prenegotiated fixed rate for each category of labor, such as, senior engineer, engineer, drafting person, typist and other personnel engaged on the contract. On many of these cost-plus contracts, the overhead charges are directly billed as a multiplier. Cost of materials is passed onto the purchaser usually with a percentage markup. Other costs such as travel, communications, reproduction, are also passed onto the purchaser at a percentage markup of the actual cost.

This arrangement eases the administration procedures for the contract since only expended time records and material invoices need to be audited for payment.

A *Cost-plus Percentage Fee Contract* is where the purchaser pays all the bills or the contractor is reimbursed for his actual costs incurred plus a specified percentage of the component costs of the work. Materials, labor, transportation, and everything else except salaries of the contractor's supervisory personnel assigned to the contract, and corporate management are generally included in the cost item of such a contract. However, the contract should clearly define those items for which the contractor will not be paid. These sometimes include his office overhead costs, including charges for any equipment or facilities that he would use to perform work under the contract.

These would then be assumed to be recovered by the fee paid.

The contractor is required to keep all records of costs and then present them to the purchaser for checking, approval and payment.

A *Cost-plus Fixed-Fee Contract* is another variation on the cost-plus contract and is where the contractor is reimbursed for all direct costs he has incurred plus a specified fixed fee.

With this arrangement, the purchaser pays a fixed fee as a remuneration to the contractor, which is usually based on elements such as the supply of certain services, overheads, and profit. It avoids the costly paperwork (to the purchaser) as well as uncertainties that may exist for the cost of labor and materials and percentages thereof to be paid to the contractor. The contractor only receives a specified sum no matter what the cost of the contract is. For this type of contract, it is only natural that the contractor will make every endeavor to expedite the work under the contract so that his personnel can be made available for another contract elsewhere as soon as possible.

It is also possible to have a contract with a lump-sum payment and provisions for other charges such as labor, materials and other cost elements (computer time, stationery, printing, reprographic services) to be made on a reimbursable basis. The purchaser is invoiced on a charge per manhour basis for each category of wage and salary labor. Material costs are passed on to the purchaser usually with a percentage markup. With this arrangement, the contractor would have lesser incentive to finish the work. It may even be to the contractor's benefit to commit surplus labor to current requirements.

There are numerous other arrangements for cost-plus reimbursable work such as the contractor's accepting an assignment to complete the work for a target cost. This type of contract aims to overcome the contractor's lack of motivation to complete the work by introducing incentives for the contractor to become as efficient as possible. Should the contractor complete the work below the target cost, he or she receives a bonus profit on an agreed proportion of difference. Should the contractor's actual costs exceed the target cost, he or she receives payment for only an agreed proportion of the excess.

It will be clear that for any form of reimbursable contract, the cost elements need to be clearly defined in the contract together with the means by which payments due to the contractor are calculated.

1.2.3   Comparison between Lump-Sum and Cost-Plus Contracts

Advantages and Disadvantages

The advantages and disadvantages between the two type of contracts are set out in Figure 1–1 shown on page 1–12.

Purchaser's Rights and Responsibilities

Figure 1–2 shown on page 1–13, compares the purchaser's rights and responsibilities between the two types of contract and concentrates mainly on matters of contract administration. While these issues are, or should be, defined in the contract conditions, experience has shown that in practice, the scope and limits of the purchaser's rights and responsibilities for each type of contract are not always understood.

It must be recognized that, unlike a cost-plus contract, for a lump-sum contract the purchaser receives the equipment and services "as bought" for the contract price and conditions set out in the contract documents. Should the purchaser require additional or alternate components of equipment and/or services during the course of the contract, the contractor is justified in negotiating a variation to the contract. Of even greater importance is that in the contract documents issued for bidding for a lump-sum contract the purchaser must clearly specify what the purchaser's contractual rights, duties, and responsibilities are. If these have been accepted by the successful bidder, then the purchaser must abide by what is contained in the contract documents.

Delays in the completion of a lump-sum contract have often been attributed to the purchaser's arguing with the contractor over exercising the right to approve certain

LUMP − SUM	COST − PLUS
GENERAL	
Complete definition of the work is essential	Requires sufficient information to enable the contractor to start with a minimum of delay
ADVANTAGES	
Allows for competitive bidding Purchaser establishes the amount of his commitment in advance and during the course of the contract Involves a minimum of administrative costs to both the purchaser and the contractor Responsibility for the performance of the contract is placed on the contractor	Requires a lesser amount of time and effort for the purchaser to prepare the bidding documents Flexibility with the purchaser's participation in the contract Minimum purchaser / contractor conflict Purchaser has some control over costs incurred Purchaser can use the contractor's resources to evaluate alternate designs Purchaser can usually terminate the contractor's services at any time without incurring substantial costs
DISADVANTAGES	
Long bidding time depending on the type and complexity of the equipment being procured Lack of flexibility makes changes to the contract difficult and expensive Costs to the purchaser may be unnecessarily high due to the contractor's contingency and risk allowances	Purchaser has less assurance of what the final contract costs will be Purchaser has to check and certify the contractor's labor, overhead, and other expense costs Contractor has less incentive to minimize the purchaser's costs on the contract

FIGURE 1−1   Advantages and disadvantages of types of contracts.

| SUBJECT MATTER | TYPE OF CONTRACT | |
	LUMP – SUM	COST – PLUS
Selection of subcontractors after award of contract	Bidder to list subcontractors in bid. List is subject to the purchaser's approval prior to award of contract. If not nominated in bid, purchaser's prior approval must be obtained.	Contractor to establish the purchaser's requirements. Purchaser's prior approval is required.
Review of contractor's proprietary design data	No	No
Review of contractor's specifications and purchase orders	No unless agreed prior to award of contract	Yes
Review of contractor's drawings and technical data	A list of drawings and technical data with submittal dates to be included in the bidding documents	Purchaser to define reviewing requirements with the contractor.
Specific inspection and testing to be performed in contractor's shop	Scope of shop testing to be clearly specified and whether certified or witnessed to be noted on equipment data sheet	Scope of shop testing to be clearly specified and whether certified or witnessed to be noted on equipment data sheet
Copies of contractor's and subcontractor's purchase orders: • priced copies • unpriced copies	Contractor to forward to the purchaser copies of all unpriced purchase orders to enable the purchaser to arrange inspection of subcontractor's work	Contractor to forward to the purchaser copies of all priced purchase orders for processing as defined in the contract documents

FIGURE 1–2   Purchaser's rights and responsibilities for types of contracts.

contractor's activities such as engineering specifications, confidential calculations, reviewing and approving purchase orders and subcontracts, and demanding additional quality control procedures over what the contractor has included in the bid. When attempting to exercise these rights, especially for the procurement of highly complex and specialized equipment, the purchaser often fails to recognize that the contractor also has his own expertise and skills that the purchaser may not always possess, but which are equally important for the successful completion of any project.

The above matters concern lump-sum contracts only. For a cost reimbursable type contract, the purchaser has the right to review and approve the contractor's calculations, design, drawings and specifications. He can also review the bid analysis and evaluation sheets and approve the recommended bidder, or can instruct that the contract be placed with another bidder before the equipment is committed to purchase.

These matters should also be read in conjunction with Figure 1–2, forming part of the advantages and disadvantages between the two types of contracts.

Liability for Faulty Work (Workmanship, Errors, and Omissions)

The contractor should always be responsible for and bear the costs of the rectification of any faulty work for a lump-sum contract. However, should the contractor undertake engineering work on a cost reimbursable basis, the cost of making good faulty work can become an exceedingly difficult area to resolve. For this reason, there should be no doubt between the parties concerning what actions are to be taken to correct any faulty work and who is to bear the cost. These matters require careful attention and must be clearly defined in the contract documents as they are directly related to the method of reimbursing the contractor for services as will be discussed in Chapter 5.

OVERVIEW OF PROCUREMENT BY COMPETITIVE BIDDING

2.0   INTRODUCTION

Due to the extremely varied scope of contracts, competitive bidding is a widely used method of obtaining and selecting contractors to undertake contracts for engineering and construction projects. However, the difference in methodology for handling contracts for the procurement of goods in the mechanical, electrical, and chemical engineering fields, compared with those for civil engineering, building and construction projects (or simply "civil works") is not always fully understood. Contracts are structured to provide for the following:

Procurement of goods. Equipment and plant systems for use in steam and power generation, mining and mineral processing, processing for petroleum, gas, petrochemical, chemical, food and beverage manufacturing facilities, and the like.

Civil works. Contracting for roads and highways, airports, ports, harbors, bridges, railways, dams, buildings, and the like. This book does not cover civil works.

Each involves a different approach to procedures to be followed for the preparation of bidding documents, evaluation and comparison of bids. Examples of this and what is entailed can be illustrated by the table below.

Activity	Procurement of Goods	Civil Works
Design responsibility	Contractor, with the contractor warranting performance features specified in the bidding documents	Owner or consultant, preparing construction specifications, drawings and bill of work which are issued with the bidding documents

(continued)

Activity	Procurement of Goods	Civil Works
Division of work	Main contractor with subcontractors as required	Much of the work undertaken by subcontractors under the control of the main contractor
Main location of work	Manufacturer's factory	Project site
Testing	Usually includes material and performance tests	Testing is usually not required other than stress and weight testing
Basis for evaluation and comparison of bids	Need not be lowest priced bid received due to possible adjustments for operating and maintenance costs	Usually the lowest priced bid that is substantially responsive to the bidding documents

This book is only concerned with presenting recommended guidelines for procurement formation procedures for the supply and delivery of major and complex equipment by competitive bidding. Competitive bidding provides the purchaser with a range of choices in selecting the best bid from competing equipment suppliers/contractors who are eligible to bid. With clear and detailed bidding procedures, they will have equal access to information, equal opportunity to bid for contracts, and receive a fair and impartial consideration of award of contract. Bidders will also be motivated to bid when they see that the ground rules are fair and clearly set forth.

2.1   STAGES OF THE COMPETITIVE BIDDING PROCESS

Figure 2–1 shown on page 2–4 summarizes the stages of the procurement formation cycle and the principal activities that take place during each stage. As each of the steps in Figure 2–1 are explained in detail in other parts of this book, the reference to the particular chapter applicable has only been shown for quick reference.

Figure 2–2 on page 2–5 shows a procedural flowchart of the procurement formation process from developing the bidders' list, including the prequalification of contractors, through to the implementation of the contract.

Stage 1.   Planning the Procurement (Chapter 4)

Success of any procurement is very much a function of having in place an effective plan specifically tailored to meet the purchaser's objectives and requirements in the procurement. In order to initiate the work, the scope of work has to be identified, as does the time by which the work must be complete, the people have to be organized and responsibilities assigned, and methods of how the work will be performed must be established. Strategic planning is considered to be a most important element of the procurement process.

Stage 2.   Preparing Bidding Documents for Inviting Bids (Chapters 5 to 8)

Following the planning stage comes the preparation of the bidding documents. This is a critically important step, since much depends on the adequacy and quality of these documents for the successful performance of the contract.

Stages 3 and 4.   Calling Bids and the Bidding Period (Chapter 9)

Calling bids is accomplished by either forwarding the invitation to bid to preselected bidders or advertising by publishing the invitation to bid in newspapers and periodicals of general circulation. The bidding period is the time between the issuance of the bidding documents for inviting bids and the date specified for bidders to have their bids received by the purchaser.

Stage 5.   Receiving, Opening, and Evaluating Bids (Chapters 10 and 11)

Once all bids have been received, bid opening may be public, restricted, or private. The evaluation process consists of appraising the technical, contractual, and financial features of the bids. Meetings with selected bidders may be necessary to discuss any qualifications, alternate bids if permitted, or other aspects not conforming to the requirements of the bid. A recommended bidder is selected and a bid evaluation report is submitted to management to seek approval to commit the procurement.

Stage 6.   Finalizing Contract Documents and Awarding the Contract (Chapter 12)

Once management has approved the selected bidder to be awarded the contract, all contract documents are updated where required by the purchaser so as to be brought to an "issued for purchase" status. The successful bidder is notified of an award of contract, reviews the contract documents, and accepts the purchaser's offer to enter into a contract. Unsuccessful bidders are then notified that the contract has been placed. The purchaser draws up the contract agreement for both parties to execute the document.

STAGE 1   PLANNING THE PROCUREMENT
- Establish the scope of the procurement
- Formulate the procurement program
- Organize people and assign responsibilities
- Develop the list of bidders [if not advertising for bids]

STAGE 2   PREPARING BIDDING DOCUMENTS FOR INVITING BIDS
- Prepare contract documents for bidding:
  - Technical documents
  - Commercial documents
- Prepare requirements for bidding
- Prepare the invitation to bid

STAGE 3   CALLING BIDS
- Issue the invitation to bid to preselected bidders, *or*
- Advertise for bids

STAGE 4   BIDDING PERIOD
- Issue supplemental notices (amendments to bidding documents) – if any
- Handling inquiries and similar situations during the bidding period

> | Bidders prepare and submit their bid |

STAGE 5   RECEIVING, OPENING, AND EVALUATING BIDS
- Register bids received
- Open bids
- Evaluate bids
- Negotiate with selected bidders
- Finalize bid analysis sheets and select recommended bidder
- Prepare the bid evaluation re/fsport for procurement approval
- Receive management's approval to commit the procurement
- Optional – Isssue letter of intent to successful bidder

STAGE 6   FINALIZING CONTRACT DOCUMENTS AND AWARDING THE CONTRACT
- Update the technical specification
- Update the commercial documents
- Update the successful bid – if required
- Assemble the contract documents for the contract
- Issue to the successful bidder:
      award of contract, *or* letter of acceptance
- Successful bidder acknowledges the award notification
- Issue notifications to unsuccessful bidders
- Successful bidder submits performance bond – if specified
- Purchaser and the [now] contractor sign the contract agreement

---

IMPLEMENTATION OF THE CONTRACT
(OUTSIDE THE SCOPE OF THIS BOOK)

FIGURE 2–1   Summary of stages of the procurement formation process by competitve bidding.

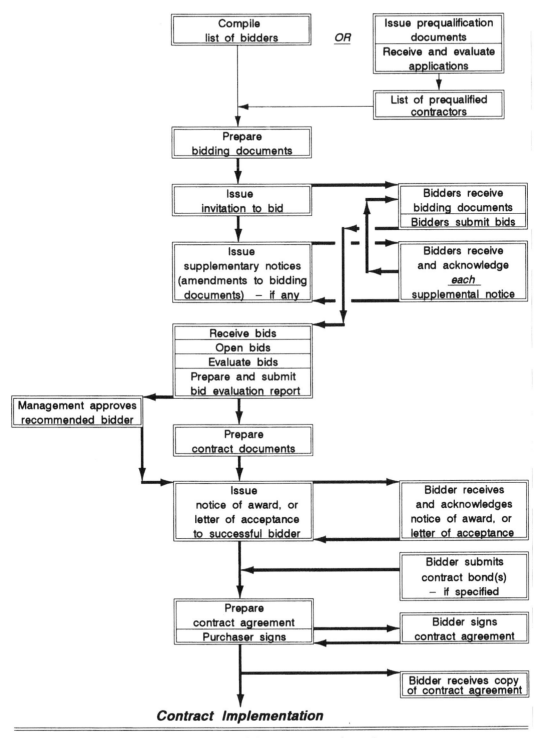

FIGURE 2–2   Flowchart of the procurement formation process.

3

THE KEY DOCUMENTS

3.0 INTRODUCTION

In Chapter 1 an engineering contract was stated to be a legally binding agreement that has been negotiated between the purchaser and the contractor for undertaking clearly specified work. The procurement formation process consists of written transactions between the purchaser and the contractor leading to a formal contract agreement to provide the specified work under the contract.

For the type of equipment procurements dealt with in this book, the formation process begins with the purchaser planning the procurement, preparing the documents for bidding, and then issuing invitations to bid for the supply of the specified equipment together with any services to be provided by the contractor. It ends when the purchaser accepts a formal offer submitted by a bidder. The contract is then formed based on the offer and the acceptance when the two parties sign the contract agreement.

The negotiation process itself can be long and involved, depending on the type, size, and complexity of the equipment being procured. In practice, it requires the preparation and issuance of a number of documents between the purchaser and bidders before a contract is formed. Unless the procurement is a repeat of identical equipment supplied previously, almost every project will require a new set of procurement documents to meet specific commercial and technical requirements.

This chapter outlines the role each key document plays in the contract formation process and also discusses the interrelationship between the documents.

3.1 STRUCTURE OF THE DOCUMENTS

From the overview of the competitive bidding process outlined in Chapter 2, a number of key documents have to be prepared for the following procurement activities:

- Bidding documents for inviting bids, *and*
- Contract documents to enable award of contract

There is no international standard that defines or provides an interpretation of terms used for inviting bids or for documents forming the contract. The same applies to names and titles where professional engineering institutions and societies, corporations, governments, and authors have designated individuals, performing the same duties and responsibilities, with different names and titles. There are also varying titles for documents serving the same function. Titles such as "bid documents," "bidding documents," and "contract documents" are in common use, but at times are applied inconsistently and with no logical uniformity. The diverse individual practices, careless use of titles, and lack of uniform order and terminology in documents have often caused considerable confusion not only to the purchaser's personnel but also to bidders and contractors.

The principal groups of procurement documents are:

- The bidding documents
- The bid
- The contract documents

The general structure of the bidding documents and the contract documents for major equipment procurements is further described below.

The Bidding Documents.   These documents form the basis of the purchaser's invitation to bid and are a group of separate documents issued to prospective bidders for inviting bids. They set out the work and services to be performed under the contract by the contractor. It is usual practice to attach to each invitation to bid a set of documents presented in three sections. Each section comprises a number of separate documents as shown in Figure 3–1 (page 3–3).

Section 1.   Bidding Requirements
- Instructions to bidders
- Content of the bid
- Bid form
- Bid security, if specified by the purchaser

Section 2.   Contract Documents for Bidding—Commercial
- Conditions of the Procurement Contract, comprising:
  - General (Standard) conditions
  - Supplementary conditions
  - Special conditions
  - Sample form of contract agreement

Section 3.   Contract Documents for Bidding—Technical
- Technical Requirements Form, comprising:
  - General information
  - A listing of technical documents

Supplemental notices form part of the bidding documents as shown in Figure 3–1. They are written notifications forwarded to all bidders before the bid closing date informing them of any amendment to one or more of the bidding documents.

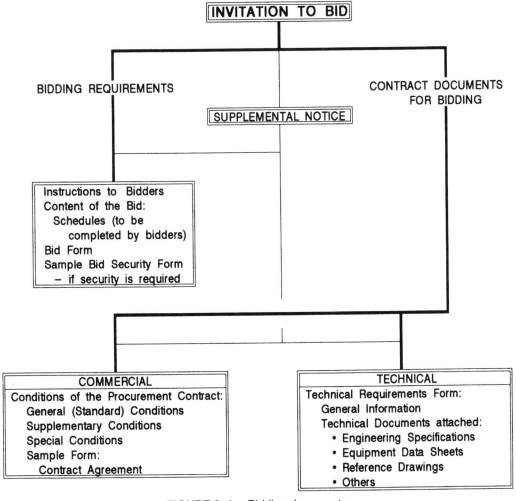

FIGURE 3–1   Bidding documents.

The Bid.   This is the bidder's offer or proposal in response to the purchaser's invitation to bid, as set out in the bidding documents. After bids have been received and evaluated, a contract is prepared as soon as a bidder has been selected to perform the work under the contract.

The Contract Documents.   Contract documents are those documents that define the contract. A contract should always specify what documents are to be regarded as contract documents. These documents are shown in Figure 3–2 (page 3–5) and normally include:

- Notice of award or letter of acceptance of bid
- The commercial documents, amended or revised following purchaser and bidder negotiations
- The technical documents, revised from bidding to an "as purchased" status
- The successful bid
- The contract agreement

It will be noted from Figure 3–2 that the instructions to bidders and the bid security have no further relevance to the procurement process after placing the contract. The successful bidder's completed schedules (see Chapter 8) and the bid form are integrated in the contract documents as are all supplementary notices and conference or telephone confirmations. The commercial and technical documents issued for bidding are carried forward to become contract documents together with the other documents shown in Figure 3–2 to form the contract.

Each of the key documents is described separately in this chapter.

3.2   REQUIREMENTS FOR BIDDING

The scope and content of the documents forming the purchaser's requirements for bidding will vary with the type, size, and complexity of the equipment being procured. The details contained in these documents should be such that bidders will fully understand the purchaser's requirements and speedily respond with the requested data and information. A clear and well defined format of instructions will enable bidders to set out their offers in a uniform manner, which will save the purchaser much time and effort when evaluating bids.

The following sections describe the scope and intent of the invitation to bid and the bidding requirements documents.

3.2.1   Invitation to Bid

An *Invitation to Bid* is a document to bring to the notice of prospective bidders the development of a new project or the acquisition of additional equipment. It describes the proposed work and services to be performed under the procurement contract. For corporate

THE CONTRACT

CONTRACT DOCUMENTS

NOTICE OF AWARD or LETTER OF ACCEPTANCE OF BID

COMMERCIAL DOCUMENTS

Conditions of the Procurement Contract:
General (Standard) Conditions
Supplementary Conditions – updated with the successful bidder's completed commercial schedules attached
Special Conditions – if any
Bid Form – completed by the successful bidder

TECHNICAL SPECIFICATION

Technical Requirements Form: – revised for purchase
General Information
Technical Documents attached:
• Engineering Specifications – revised for purchase
• Equipment Data Sheets – revised for purchase
• Reference Drawings – revised if required
• Others – revised if requred

SUCCESSFUL BID – with completed technical schedules

CONTRACT AGREEMENT

CONTRACT SECURITIES – if specified

BIDDING DOCUMENTS

INVITATION TO BID

BIDDING REQUIREMENTS

Instructions to Bidders
Content of the Bid – Schedules for bidder completion
Bid Form
Sample Bid Security Form – if security is required

COMMERCIAL DOCUMENTS

Conditions of the Procurement Contract:
General (Standard) Conditions
Supplementary Conditions
Special Conditions – if any
Sample Form – Contract Agreement

TECHNICAL SPECIFICATION

Technical Requirements Form:
General Information
Technical Documents attached:
• Engineering Specifications
• Equipment Data Sheets
• Reference Drawings
• Others

SUPPLEMENTAL NOTICE – if any

FIGURE 3–2   Bidding documents and contract documents forming the contract.

3 – 5

works, the bid invitation together with the bidding document package are normally forwarded to bidders whom the purchaser knows are both technically and commercially competent to undertake the work. An alternate method of inviting bids, and one that is used mainly for public funded projects and those funded by major development banks, is through published notices and advertisements appearing in newspapers and appropriate trade journals. The purpose of the latter is to attract as many prospective bidders as possible who may be interested to bid for the work.

3.2.2   Bidding Requirements

The *Bidding Requirements* documents, also known as the Conditions of the Bid, are the "rules" set by the purchaser that the bidders are to comply with during the bidding period. Bidders should be made aware that the role of these documents is to set forth the criteria and basis under which their bids are to be prepared, submitted, received, and evaluated.

Three separate sets of documents cover the bidding requirements:

1. Instructions to bidders
2. Content of the bid
3. Bid form

Instructions to Bidders

As the name implies, the *Instructions to Bidders* document informs all bidders of the purchaser's precise requirements for the preparation and submission of their bids. The instructions also include the general information and data that each bidder should be familiar with in order to prepare a bid.

Detailed procedures for the preparation of this document are discussed in Chapter 8. It is important to recognize that the instructions to bidders document should contain only matters relating to the preparation and submission of the bids. Once bids have been submitted and received by the purchaser, this document has served its purpose and will not form part of the contract.

Content of the Bid

The *Content of the Bid* specifies the purchaser's precise technical, contractual, and financial information and data to be provided by the bidders. The details required are presented in the form of a series of schedules to be completed by bidders.

The term "Schedule" appearing in bidding and contract documents has often been used loosely and then quite inconsistently applied in the documents. Schedules can be used three ways in documentation:

1. In the form of questionnaires prepared and issued by the purchaser with the bidding documents and required to be completed by the bidders
2. A timetable, plan, or program for engineering, procurement, and construction for all or part sections of the work under the contract
3. A listing of items prepared by either the purchaser or contractor

In the interest of consistency, the term "Schedule" in this book refers to the first item above—for commercial and technical information requested by the purchaser to be completed by the bidders and submitted with their bids (see Chapter 8, page 8–8).

When preparing the contract, the successful bidder's completed schedules become part of the formal contract documents as shown in Figure 3–2. The commercial schedules will form part of the supplementary conditions (see Paragraph 3.3.1), while the technical schedules are included in the successful bidder's bid.

Bid Form

The *Bid Form* is a document prepared by, and addressed to, the purchaser in which the bidder states a price to perform the work under the contract and to supply other information sought by the purchaser. When completed and signed by the bidder, it is legally deemed to be an offer by the bidder to enter into a contract with the purchaser on the conditions stated in the bid form. The same bid form is issued to all bidders so that the uniform arrangement of data to be furnished will avoid bidders submitting offers in different forms or with incomplete information, contingencies, or superfluous data. It also facilitates the bid comparison and evaluation process. Bid forms frequently used are preprinted standard forms published by a number of national engineering institutions. Some private companies and government agencies may prefer to use in-house bid forms specially tailored to meet their requirements.

3.3   COMMERCIAL DOCUMENTS

In simple terms, the commercial section of a procurement covers everything that is not a technical matter. Contrary to common belief that the term "commercial" is solely related to price, delivery, and the like, it also covers the financial and legal aspects of the procurement. Apart from the previously discussed bidding requirements, the commercial requirements are set out in the conditions of the procurement contract as follows.

3.3.1   Conditions of the Procurement Contract

This book is only concerned with written contracts, since oral or spoken negotiations and agreements often lead to misunderstandings and disputes arising during the execution of the contract. To prevent such occurrences, the purchaser prepares a set of written "rules" that the two parties agree on prior to signing the contract agreement. The rules concern their rights, obligations, and responsibilities in the way the contract is to be run and are commonly termed Conditions of the Contract. For a procurement contract, they are referred to as the *Conditions of the Procurement Contract*, or simply, Contract Conditions.

Not only do the contract conditions define the purchaser's and contractor's rights, obligations, and responsibilities for administrating the contract, they also set out what course of action is to be taken in the event of either party failing to fulfil their contractual obligations.

The nature and extent of the contract condition documents will depend on such factors as the size and complexity of the equipment being procured, as well as an assessment of situations and conditions that may be encountered. The following two or three separate sets of documents form the contract conditions:

1. General (Standard) conditions
2. Supplementary conditions
   or
1A. One set of contract conditions incorporating both 1 and 2
   and
3. Special conditions

This book deals only with the three documents (1, 2, and 3) as option 1A will vary with each purchaser's operating and business activities.

General (Standard) Conditions

General (Standard) Conditions are those conditions that are applicable to most contracts undertaken. A number of standard or model forms of general conditions have been prepared and published by national engineering institutions and standards association societies. The wording in these documents has been carefully chosen and vetted by legal experts in contract law, as the conditions can have considerable legal implications and consequences in their application. It is suggested that copies be obtained for the purpose of analyzing their suitability for the purchaser's particular project.

General contract conditions cover such matters as:

- Applicable law
- Contract language (English or other)
- Subcontracts and assignment
- Patent rights
- Force majeure
- Termination for default, insolvency, and convenience
- Resolution of disputes

Supplementary Conditions

The general (standard) conditions are expected to be met in most contracts undertaken. In practice, the contract conditions must cover an extremely wide range of procurements and as a consequence, there is usually a need to draft additional or supplementary conditions to meet the contractual requirements for a particular contract undertaken. These additional contract conditions are generally set out and issued in a separate document titled *Supplementary Conditions* and contain such matters as:

- Variation, inferring a modification to the contract arising from a change to the scope of work as provided for in the contract, but not always altering the contract price
- Equipment warranty or defects liability
- Contract program showing milestone completion dates
- Delivery
- Extension of time to the contract
- Taking over, certification, and payments
- Procedures on technical matters such as inspection, progress reporting, and work program

Included after award of contract are the successful bidder's commercial schedules.

Although the commercial section of a contract usually incorporates both the legal and financial aspects of a contract, the importance of identifying certain key financial issues relating to a contract is often overlooked. It is one area in the procurement where both the purchaser and the contractor should be closely concerned, as it is here, depending on how successful the contract is, that large sums of money can be lost or saved. It is essential, therefore, that these financial matters be clearly defined. They include, but should not necessarily be limited to, the following:

- Contract price structure
- Contract price adjustments due to fluctuations in currencies, cost of labor, materials, or transportation during the course of the contract

- Terms of payment
- Liquidated damages
- Insurance

Special Conditions

Special Conditions are sets of individual documents classified as company policies, or state or federal government acts and regulations that would apply to the particular procurement. These conditions cover policies on the employment of apprentices, purchasing preferences, industrial agreements, and the like.

3.4   TECHNICAL SPECIFICATION

It has often been stated that contract conditions vary little from one contract to the next with most provisions remaining reasonably standard. Although this is generally true for commercial conditions, it is not the case for the technical aspects of major equipment procurements as the work can cover a wide range of variables to meet the purchaser's design, operating, and maintenance requirements.

Unfortunately, the term "specification" is extensively used to infer only technical matters. This is not correct as a specification by definition is a statement of requirement, as it may cover not only technical issues but also commercial or any other requirements. In this book, a document covering a set of specific technical requirements has been termed an "engineering specification." Usually there are a number of engineering specifications covering different sections of the work. All the technical requirements for a procurement form the *Technical Specification.*

The scope of work covered by the technical specification is summarized in the technical requirements form, as described in the next section.

3.4.1   Technical Requirements Form

The *Technical Requirements Form* contains the technical data and requirements for a equipment procurement and incorporates the following information:

1. General Information

   - Name/title of the equipment being procured
   - Quantity and description of equipment being procured
   - Contract number
   - Destination where the equipment is to be delivered
   - The current revision status of the document

2. A list of all the technical documents forming the technical specification each with the current revision status. Normally included are:

- Engineering specifications (a written document)
- Equipment data sheet(s)
- Reference drawings
- Other technical documents that are relevant to the procurement

Because the current status of the technical requirements form is updated each time a change is made to either the equipment quantity or an amendment to one of the listed documents, the form becomes a checklist of information for bidding purposes.

The form is reissued when the equipment is committed for purchase and becomes one of the contract documents. Prior to reissuance, the status of the form is revised from "issued for bids" to "issued for purchase" with all the listed documents brought to an "as purchased" revision.

The form remains a working document until completion of the contract. The distribution of the form will ensure that the contractor as well as the purchaser's personnel assigned to the procurement—including engineers, inspectors, expediters, and construction personnel—are kept informed of the latest revision of each document in the technical specification until the contract is complete.

3.5   SUPPLEMENTAL NOTICES

A *Supplemental Notice* is a written notification forwarded to all bidders by the purchaser prior to the bid closing date advising of an amendment by way of a correction, addition, deletion, or clarification of information to a bidding document. Figure 3–1 shows the function of the notice in relation to the other bidding documents.

Supplemental notices may also be used where the purchaser is not in a position to issue a complete set of bidding documents but wishes to release sections of the documents to bidders to get the project started. This up-front bidding information is forwarded to the bidders through means of the supplemental notices. Confirmation that the information detailed with each supplemental notice has been incorporated in the bid by the bidder must always be noted in the bid form.

3.6   SUCCESSFUL BID

The *Bid* is the bidder's formal offer submitted in response to the purchaser's invitation to bid. The bid should be prepared in complete conformance with the bidding documents. As this book is concerned with competitive bidding, bids will have been received from those bidders who accepted the bid invitation and be fully evaluated for the selection of an offer that will best suit the purchaser's requirements for the procurement. After bid evaluation, the successful bid becomes a contract document, as shown in Figure 3–2.

3.7   NOTICE OF AWARD—LETTER OF ACCEPTANCE

Following the completion of the bid evaluation and management's approval (as appropriate) to commit the procurement, the purchaser issues a formal notification to the successful bidder of the acceptance of the bid. The document is usually termed as the *Notice of Award* or *Letter of Acceptance of the Bid* and is an acceptance of the bid. At this stage, the acceptance of the bid forms a binding contract between the purchaser and the bidder, pending the signing of a formal contract agreement as provided in the bidding documents.

The notification should clearly state the basis of acceptance, and the date of acceptance of the bid. As the commercial and technical documents and the successful bid will now have been brought to an "as purchased" status as shown in Figure 3–2 (also see Chapter 12), it is prudent to attach a full set of contract documents with the notice to allow the successful bidder the opportunity to agree that this is indeed the contract that he has entered into, and to allay any fear of errors having crept into the contract. Attaching the contract documents will of course waylay any arguments, legally or otherwise, in the administration of the contract at a later date as the contractor has agreed to their acceptability.

Once the bidder has acknowledged receipt of the notification, both the purchaser and the contractor are then obliged to sign the contract agreement.

The date of the notification of acceptance is when the contract comes legally into existence and is usually called the "contract date." Unless specified otherwise, this is Day 1 of the contract and therefore has considerable contractual significance.

The award of contract should not be confused with a notice or letter of intent. This notice solely infers that it is the purchaser's intention to place a contract with the selected bidder, and allows him sufficient time to finalize what may be a complicated set of contract documents while allowing the contractor time to organize a project team to perform the work under the contract.

There may also be situations when the purchaser has not finalized all the contract documentation but wishes the work to commence immediately because of a tight schedule. In this instance, the purchaser issues a notice or instruction to proceed advising the successful bidder, in a sparsely worded manner, of the acceptance of his bid. This notification provides an instruction as to when to start the work or the specified part of the work, conditions, and methods for payment for part work completed, and advises that a notice of award will be forthcoming on completion of all documentation.

Certain countries have defined rulings as to the meaning of "to proceed" and how the document should be worded, so this should be checked for legal standing. The contractor may not wish to proceed with the work until a formal acceptance of the bid is received, which must be accepted by the purchaser as the contractor's right. This, therefore, should be checked for legal standing.

3.8   CONTRACT AGREEMENT

The *Contract Agreement* is a written agreement, enforceable by law between the signatory parties. The purpose of the contract agreement is to identify the conditions under which the equipment and services are to be furnished by the contractor, in consideration of the payments to be made to the contractor by the purchaser.

The contract agreement itself is not the entire contract. It is only one of the contract documents that make up the contract as shown in Figure 3–2. Nevertheless, since it references all other documents, the contract agreement is a very important part of the set of contract documents as it is the one to which the contracting parties affix their signatures and has special standing in the eyes of the law.

Various forms of contract agreements are often included in the general (standard) contract conditions published by engineering institutions and standards associations. These forms are brief and contain fill-in spaces to be completed by the purchaser. Each preprinted agreement should be carefully reviewed for each new contract undertaken to ensure it meets all conditions and requirements. Most of the larger corporations, as well as government agencies, have their own standard forms of contract agreement.

In order to minimize the time-consuming efforts for reviewing the final documentation after awarding the contract, it is good practice to include a copy of the contract agreement form with the bidding documents. It is not only a matter of courtesy but also smooths final discussions, and shows bidders the document first rather than causing a possible surprise after precontractual negotiations have been concluded. Where a standard form is used, blank spaces should be completed wherever possible so that bidders will not be confronted with important data that will be seen for the first time just before signing the agreement.

3.9   BID AND CONTRACT SECURITIES

For major equipment procurements, the purchaser may seek a means of protecting his interests should unexpected eventualities arise, such as the selected bidder failing to accept the contract as awarded to him, or after award of the contract, the contractor defaults on performing the obligations as defined in the contract documents. Such protection is accomplished by the bidder/contractor providing the purchaser with a bond as a form of security that will guarantee that he will fulfill his obligations as defined in the bidding documents. Each bond is backed by a monetary amount.

There are three types of bonds in common use in engineering contracts. They are:

Bid Bond. Provided by the bidder as an assurance not to withdraw the bid and will sign the contract if it is offered.

Performance Bond. Provided by the contractor to whom the contract is awarded. The bond makes certain that, should the contractor fail to perform any of the terms and conditions of this contractual undertaking, the purchaser is protected against loss up to the pecuniary amount of bond penalty.

Payment Bond, also called *Labor and Materials Payment Bond.* Provided by the contractor to whom the contract is awarded as a guarantee that payment will be made for all the labor and materials used, including payment to the contractor's subcontractors, material suppliers, and protects the purchaser against loss should the contractor fail to do so.

Bonds are normally completed and signed on preprinted forms. Unless they are carefully worded, and the conditions clearly defined, they could become worthless documents. Care must therefore be taken that the bonds are properly prepared and properly executed.

Bid and contract securities are discussed in more detail in Chapter 13.

PART TWO

PROCEDURES FOR PROCUREMENT FORMATION

This part of the book, covering Chapters 4 to 12, is designed to show procedures for setting up a major equipment procurement. It aims to guide the engineer through the entire procurement formation process with a step-by-step description of each stage from planning the procurement through to the signing of the contract agreement. Meticulous planning is essential in order to ensure that the contract will be completed within budget and on time. The process then proceeds with the preparation of key documents that define the requirements for the work; developing the commercial aspects of the contract and writing the technical specification; preparing invitations to bid, receiving and evaluating bids; and finally negotiating a contract agreement to be executed by the purchaser and the successful bidder. Discussion on implementing the contract is outside the scope and purpose of this book.

It is important to recognize that no hard and fast rules can be applied to this work. The information presented is for guidance only. Each procurement should be considered on its own merit and the engineer is expected to act prudently in adopting the applicable requirements to each particular project.

The reader's attention is drawn to Part 5, which shows how key bidding and contract documents are prepared and presented in practice for a major equipment procurement.

4

STRATEGIC PLANNING PROCEDURES

4.0 INTRODUCTION

The successful implementation and completion of any major equipment procurement is essentially a function of realistic planning tailored to achieve the purchaser's project objectives. The importance of effective planning is to ensure that the bidding process, and later, administrating the contract, are totally manageable. This has not always been recognized, as all too often insufficient time and effort have been given to this section of the work. It is here where much of the purchaser's money, time, and effort can most easily be lost or saved.

Planning must proceed in logical and sequential steps. However, before the process can commence, it is essential that the people responsible for developing these activities have a clear understanding of the work to be performed. It is also assumed the following have been completed to establish the viability of the procurement:

- Conceptual engineering and design criteria to finalize equipment design and performance parameters, as well as plant location and layout
- Review of various equipment options leading to the one most suited for the proposed service
- Preliminary delivery times obtained from reputable equipment suppliers together with estimated times for installing and commissioning the equipment
- Budget cost estimates of the work based on prices obtained from reputable equipment suppliers together with in-house estimates for other work
- Statutory permits and approvals for the work to proceed
- Arrangements for financing the project

This chapter describes the procedures and provides guidelines for planning the procurement in the following sequence:

1. Identifying the procurement and what has to be achieved
2. Establishing a program in terms of time to complete the procurement

3. Organizing people and allocating responsibilities
4. Developing the procurement criteria
5. Identifying and organizing the data
6. Determining the method for bidding
7. In-house filing and document distribution systems

4.1   IDENTIFYING THE PROCUREMENT

Each procurement should be identified by the name and address of the purchaser, the title of the project, the project number, the contract number, the name of the equipment procured, and the applicable equipment tag number. This information is needed for general administration, correspondence, technical documents such as specifications, data sheets, drawings, certification records, for filing, and accounting purposes, and will ensure the efficient handling of the procurement with the minimum chance of any confusion.

4.1.1   Title and Contract Number

The following information could apply to an equipment procurement:

Name of the purchaser	—	XYZ CORPORATION
Address of the purchaser	—	2547 Tenth Street
		Anytown, NJ 00001
Title of the project	—	Devin River Mine Project
Project number	—	1712
Contract number	—	C – 1363
Equipment title	—	Fire Water Pumps
Equipment tag numbers	—	1201 and 1202

Correspondence

As the purchaser will be involved in a considerable amount of contact with bidders, and later with the contractor for any major procurement, the most responsible means of communication is by written words. It is important therefore, to establish a format to be used whether by mail or by facsimile. While layout of external correspondence usually follows in-house procedures, all correspondence to bidders should be written on the purchaser's company letterhead, with the date of the correspondence and the purchaser's reference number, and be addressed to the person with correct title. The main block should show the title of the purchaser's project, the contract number, and the subject equipment. An example of the format of such letter follows.

Example of a Letter from Purchaser to Bidder

XYZ CORPORATION
2547 Tenth Street
ANYTOWN NJ 00001

Our Ref. B/225
March 30, 1992

Buldina Pump Co.
6635 Prince Ave.
Jamesville, PA 10000

Attention: Mr. D.L. Shearer
              National Sales Manager

Gentlemen:

Devin River Mine Project
Contract No C-1363, Fire Water Pumps, Tag Nos. 1201 and 1202.

Attached are three (3) copies of Written Confirmation No. 1 in connection with the telephone conversation the writer had with your Mr. David Shearer when agreement was reached to bring forward the delivery of the above pumps and a revised contract price as noted.

Sincerely,
XYZ CORPORATION

_____
T.W. Maybolt, Jr.
Project Manager

Attachment

4.1.2   Cost Coding and Equipment Tag Numbers

The purpose of *Cost Coding* the equipment is to identify costs for each item of equipment. The cost code assigned can also provide a quick recognizable reference for procurement, specifications, equipment data sheets, filing, and for numerous other uses.

The following is an example of a series of cost coding equipment:

1100	Generating Plant Equipment
1200	Pumps
1300	Furnaces and Fired Heaters
1400	Heat Exchange Equipment
1500	Compressors, Fans
1600	Pressure Vessels
1700	Storage Tanks, Bins
1800	Stacks
1900	Environmental
2000	Size Reduction (Crushers, Mills, Rock Breakers)
2100	Classification (Screens, Cyclones, Thickeners)
2200	Conveying (Feeders, Conveyors)
2300	Dust Collection
2400	Measurement (Weighers, Belt Scales, Samplers)
2500	Agitators
2600	Cranes, Hoists
2700	Ducting, Chutes
3000	HV/OH Reticulation
3100	Transformers
3200	Switchgear, MCC, Distribution Boards
3300	Motors
3400	Cabling, Connections, Lighting, and Grounding
3500	Cable Trays
3600	Communications
4000	Piping
4500	Valves and Fittings
5000	Insulation
5100	Noise
5600	Painting
6000	Instrumentation

Other cost codes are allocated as required.

Examples could be:
Cost code 1200 for pumps and drivers for a particular project. They could include:

- Fire water pumps
- Dewatering pumps
- Bore water pumps

The cost codes for procurement could then be:

Cost code 1201—for the procurement of the fire water pumps
Cost code 1202—for the procurement of the dewatering pumps
Cost code 1203—for the procurement of the bore water pumps

An *Equipment Tag Number* is a unique identification, usually assigned by the engineering department, that physically is a metal tag attached to the item of equipment or etched on the nameplate. The equipment tag number also provides a reference for the equipment on drawings, on-site location, maintenance schedule, and history.

An example of equipment tag numbers could be:

Equipment tag no. 1201 is the first fire water pump within the plant area or system/line/circuit
Equipment tag no. 1202 is the second fire water pump, and so on

4.2   ESTABLISHING THE PROCUREMENT PROGRAM

Based on the date when the equipment has to be in continuous commercial operation, establishing the procurement program for milestone dates must first start with determining the major tasks and activities, indicating how each task will be accomplished and managed, and identifying the personnel necessary to carry out the various activities. The proposed start and completion dates for each activity must be clearly indicated. Critical dates to be established will include:

At Home Office

- Deadline for completing the bidding documents
- Deadline for issuing invitations for bids
- Bid closing date, and length of bidding period, including validity period of bids

- Deadline for the completion of the bid evaluation
- Deadline for submitting the bid evaluation and recommendation report—if one is required
- Deadline for awarding the contract (becomes Day 1 of the contract)

Without the establishment of these critical dates, confusion and delays can make a contract unworkable and cause cost overruns.

On the Site

- Deadline for receiving equipment components at the project site
- Deadline for the completion of the installation and transfer of responsibility to operating and maintenance personnel for commissioning the equipment
- Deadline for the commencement of uninterrupted operation

Figure 4–1 on page 4–15 is an example of a form used for developing the procurement program and shows the main activity items, document preparation responsibilities, and timing to ensure documents such as requesting bids, and for placing the contract are complete by the programmed dates.

4.3   ORGANIZING PEOPLE AND ALLOCATING RESPONSIBILITIES

As soon as the procurement program dates have been established, the purchaser has to organize a precontracts team and allocate responsibilities. The composition of the team will obviously depend on the size and complexity of the procurement. For smaller contracts, one person may be responsible for more than one activity, with certain commercial activities being the responsibility of the purchasing department. However, some companies, because of their size and nature of their business, do not have a purchasing section, yet take on the procurement of highly capitalized equipment. In such cases, the engineers also become responsible for the commercial aspects of the contract. Whether to bring in outside assistance will depend on the purchaser's in-house resources. Hasty decision-making by management to initiate the planning process, and to seek outside help without carefully evaluating what, why, and when it is needed and who is to be appointed, can often lead to misunderstandings, waste much valuable time, become extremely costly, and ultimately get the procurement off to the worst possible start.

As to the composition of the team, the purchaser must first appoint a person, frequently a project manager, to assume total responsibility for executing all stages of procurement formation process. After the formal contract agreement has been signed, the same person may not necessarily be the individual responsible for implementing the contract. The appointment of the procurement formation project manager is a critical factor in the success of the contract. A good project manager should possess strong human relations and leadership skills, sound knowledge of project management principles and techniques, the ability to exercise effective control in difficult and complex situations, a sense of fairness, the capability and experience to arbitrate as well as compromise, and the willingness to make and take responsibility for decisions.

The project manager normally should have the following support team:

- *A contracts manager (in larger organizations)*—responsible for compiling the commercial terms and conditions, and assembly of bidding and contract documentation
- *The project engineer*—for coordinating the technical aspects of the contract. On smaller procurements, this person may also be the project manager
- *Engineers*—responsible for writing specifications, and evaluating the technical features of bids
- *A procurement supervisor*—for purchasing activities, such as inviting and receiving bids, assisting with the commercial evaluation of bids, inspecting the proposed manufacturer's facilities, and placing the contract
- *A program planner*—for planning the procurement program
- *A cost controller*—for monitoring cost estimates and expenditures
- *A construction manager*—for the installation and assisting with the commissioning of the equipment
- *Accounting department*—for allocating costs and attending to general office administration
- *A legal department representative*—for reviewing the legal implications of the contract

Additional assistance is usually provided by:

- The plant production department for process input data and commissioning and operating the equipment. For mining equipment, this person may be the geologist.
- The plant maintenance department for review of maintenance and spare parts requirements.

Figure 4–2 on page 4–16 shows a typical organization chart of the purchaser's project team assigned to the contract and the reporting relationship with the contractor for a major equipment procurement.

4.4   DEVELOPING THE PROCUREMENT CRITERIA

For major procurements, some research and consideration must be given as to the criteria forming the basis of the equipment design, transportation, and installation before preparing the specification for bidding purposes. Such criteria would include:

- *Fabrication*—benefits of shop fabrication as compared with on-site fabrication, and evaluating the suitability of modular construction
- *Site environmental conditions*—maximum and minimum temperature, prevailing wind direction, relative humidity, hurricane or tornado frequency, storms, monsoons, rainfall, snowfall, frost line, flooding, earthquakes, dust, wind, seaboard, corrosion (from chemical vapors), and so on
- *Shipping conditions and facilities*—winter freeze, harbor tidal conditions, water depth, wharf handling capacities, barge cranes
- *Transportation*—effect of government transportation regulations, general traffic, land, air, or existing waterway transportation, road width, road clearance, bridge clearance, maximum bridge load allowance, rail bridge clearance, overhead power line clearance, en route flooding, and site off-loading facilities
- *Installation and infrastructure*—local standards and regulations, language and communication problems, experience of construction management and supervision (depending on size and complexity of equipment), water requirements and supply availability, power requirements and supply, local procurement of labor, equipment, and material, local labor expertise, existing skills and trades, union constraints, and industrial relations
- *Operating and maintenance*—quality and experience of supervision and personnel, existing skills and trades, training facilities, and industrial constraints, for example, the use of certain tools
- *Insurance*—all risks relevant to the procurement

4.5   IDENTIFYING AND ORGANIZING THE DATA

For the purpose of identifying the key data to be included in the technical and commercial documentations, the task can be broken into four major steps:

1. What are the main purposes of the bidding, commercial, and technical documents?
2. Who will use the commercial and technical documents?
3. What information does each user need to know, why and when?
4. What key provisions are to be included in the documents?

4.5.1   What Are the Main Purposes of the Bidding, Commercial, and Technical Documents?

The purpose of these documents is to provide information not only to prospective bidders but also to the purchaser's personnel.

- *Bidding*—to provide the basis for the method of bidding for the contract
- *Commercial*—to provide the terms and conditions of the contract
- *Technical*—to provide equipment design and performance requirements, and other objectives to be met

4.5.2   Who Will Use the Documents?

When planning the contents of the commercial and technical documents, it remains important to identify who will use the documents. The bidding documents will be used by companies bidding for the procurement, and by the purchaser's personnel. The commercial documents will be used by the contractor and the purchaser's contract administration and financial staff. The technical documents will be used by the contractor and the purchaser's technical staff assigned to the contract.

4.5.3   What Information Does Each User Need to Know, Why, and When?

In the planning process, it is necessary to establish what information each user, purchaser, bidder, and after award of contract, the contractor, needs to know, and how the person(s) will use that information, as well as when it will be needed. The "what and how" are the basics of requirements, and "why," and "when" will help to avoid logistical problems. It is important to ensure that all users will receive the information they need when they need it. The foregoing is outlined in Table 4–1.

Table 4–1   What Does Each User Need to Know, Why, and When?

User and What Information	Why and When
Purchaser's Purchasing Supervisor: Procurement Program	For issuing, expediting and receiving bids
Bidders' bids	After receiving bids, review them

(continued)

Table 4–1   (Continued)

User and What Information	Why and When
Purchaser's Engineer: General design, performance, warranty requirements	To write the technical specification
Bidders' bids	After receiving bids, review them
Contractor's design, calculations, and fabrication drawings	Check after award of contract
Bidders: Bidding, technical, and commercial requirements, to determine whether to bid; to prepare bid package	After receiving bids
Purchaser's Contracts Manager: Commercial requirements	Write the commercial documents
Bidding documents	Review bidding documents prior to inviting bids
Bidder's bids	After receiving bids, review bid analysis
Contract agreement	Check document before award the contract
Contractor's Engineer: Technical specification to perform calculations, design, and detailed drawings	Supervise the work after award of contract
Inspector—Purchaser or Third Party: Technical specification and shipping requirements for: • Contractor's documents • Contractor's compliance for fabrication, and delivery • Contractor's compliance with shipping requirements	Check after award of contract

4.5.4   What Key Provisions Are to Be Included in the Documents?

The final step for identifying and organizing the data before commencing to prepare the documents is to determine what key provisions must be included in the bidding, technical and commercial documents to meet each user's need. The following is a list of provisions usually used for equipment procurements. In compiling these provisions, care must be taken to ensure that they interface and do not conflict.

- *Bidding*
  - Instructions to bidders
  - Content of the bid (what bidders are to include in their bid)
- *Commercial*
  - General conditions of the contract (standard provisions)
  - Specific contract conditions for the particular procurement
  - Contract program
  - Contract price structure
  - Fixed price or prices subject to escalation
  - Terms of payment
  - Method of payment
  - Time for delivery, including bonus and/or penalty if applicable
  - Method of transportation
  - Insurance
- *Technical*
  - Scope, purpose, limits, exclusions
  - Reference documents, codes, standards
  - Operating and performance
  - Equipment selection criteria
  - Design
  - Material selection
  - Fabrication
  - Quality assurance, testing, inspection
  - Performance and defect warranties
  - Surface preparation, painting
  - Insulation
  - Shipping (special requirements)
  - Spare parts

4.6   DETERMINING THE METHOD OF BIDDING

An important part of the planning process is determining by which method contractors are to be chosen. As stated in Chapter 2, since this book is only concerned with competitive bidding, the two most commonly used methods are:

1. Open unrestricted bidding by advertisement
2. Bidding from a selected list of bidders known to the purchaser to be qualified to undertake the work

An alternate method for preparing a list of bidders, one that is sometimes used, is for the purchaser to invite contractor/equipment suppliers to first apply for prequalification as bidders to undertake the work.

4.6.1   Advertising for Bids

Advertising for bids in newspapers and journals is usually aimed at interesting as many contractors as possible in the proposed work to secure the benefit of keen competitive bidding. A published advertisement is a very effective way of reaching prospective bidders who might otherwise fail to learn about the project.

In many countries, it is mandatory that federal, state and municipal contracts for public-funded projects be advertised. This type of open bidding can attract a large number of bids and it is not uncommon for no less than 15 bidders to respond to the bid invitation, all bids of which should be evaluated. Much criticism has been voiced at this bidding process. There can be bidders who have little interest in securing the contract, and/or others who are not sufficiently competent to undertake the work. Another reluctance is that the purchaser frequently has to accept a bid from anyone who wished to compete for the contract. As a result, the purchaser may eventually lose control of the bidding process by not knowing in advance the contractors who are bidding until all bids have been received. This can involve the purchaser in unnecessary expenditure, and be a very time-consuming burden when evaluating bids.

Open bidding can also discourage many reputable or suitable bidders from bidding, as they become reluctant to spend the time, money, and effort when the competition would be too great. There is then the possibility of an unrealistic low price or "cut price" bid being submitted and accepted. The situation becomes worse for the purchaser should the bidder ultimately prove to be unsuited to undertake a complex procurement.

Some international financial institutions such as The World Bank and the Asian Development Bank, where the proceeds from loans are used to finance the contracts, may require borrowers to advertise bid invitations on an international basis. Other things being equal, preference is usually given to member countries.

4.6.2   Developing the List of Bidders

The alternate method and one that most corporate organizations follow is to prepare a list of contractors or equipment suppliers whom the purchaser knows from past experience are reliable for performance, technically competent, and have the financial resources to complete the particular work under the contract. Contractors who have previously furnished procurements to the purchaser's satisfaction and who are familiar with the purchaser's procedures would be favored bidders. Because of the high costs incurred in the preparation of bidding documents, evaluating and negotiating bids, the list should be short, limited to perhaps approximately eight organizations, and certainly not less than three, that are judged to be best qualified for the procurement.

Where advertising is not mandatory for public-funded contracts, the authorities will use a list of bidders who are known to them. In such cases, the authorities are likely to have a very elaborate procedure in place for screening prospective bidders to ensure fairness and equal opportunity in order to justify their choice, and to avoid public criticism of their selection.

4.6.3   Invitation to Prequalify for Bidding

Another method for developing the bidders list is to analyze the qualifications of potential bidders before issuing bid invitations. This is to ensure that only technically and financially capable firms will be invited to submit bids. The procedure involves the purchaser inviting interested contractors/equipment suppliers to apply for prequalification as bidders. The invitation is issued either by letter, or advertising in national—and where applicable, overseas—newspapers.

The letter or advertisement outlines the project and contract in general terms, and states the closing date for qualifications to be submitted by bidders who propose to bid for the contract. Each response is evaluated having regard to the ability of the interested bidder to perform the work satisfactorily, taking into account experience and past performance on similar contracts, capabilities with respect to personnel and manufacturing facilities, and financial position. The purchaser then selects the most qualified firms (three to 10) who will be notified that they will be issued with a formal invitation to bid.

The following may serve as a checklist for analyzing and evaluating the applications:

- Type of company
- Parent company or associated companies
- Scope of business
- How long established
- Number of employees employed

- Average annual turnover
- Management sufficiency and purchaser/contractor accessibility
- Contractor's engineering design, manufacturing, and quality control qualifications
- Proven reliability to meet delivery programs
- Record of personnel relations, strikes, and labor unrest
- Financial stability to meet warranties
- Current and future workload and ability to meet purchaser's contract program
- Whether the contractor has previously supplied equipment to purchaser
- Access to after-sales service and availability of spare parts

4.7   IN-HOUSE FILING AND DOCUMENT DISTRIBUTION SYSTEMS

It is essential that the purchaser initiates and maintains an organized filing system right from the onset of the procurement, so that all information relating to the procurement is immediately available when required. Nothing can be more frustrating and time-wasting than scouting round in numerous locations looking for vital information needed for immediate action. The filing system could be set up on instructions from the project manager but normally follows a standard in-house procedure.

   It is equally important to ensure that all who are assigned to the purchaser's procurement formation team receive copies of documents that directly concern them in the execution of their respective duties. The recommended method of overcoming "I was never informed" is for the project manager to prepare an *In-house Document Distribution Form* an example of which is shown in Figure 4–3 on page 4–17. After distribution, every original document should be placed and kept in the respective contract record file.

EXAMPLES OF RELEVANT DOCUMENTS

The following documents are included at the end of this chapter:

- Figure 4–1   Example of a bidding and contract document formation form
- Figure 4–2   Example of purchaser and contractor reporting relationships for a major equipment procurement
- Figure 4–3   Example of an in-house document distribution form

XYZ CORPORATION		
BIDDING AND CONTRACT DOCUMENT FORMATION FORM		
PROJECT _____	CONTRACT No _____	
DOCUMENT AND ACTION	PERSON / DEPT RESPONSIBLE	COMPLETION DATE
1. LIST OF BIDDERS • Prepare preliminary list of bidders for approval, or advertise for prequalification • Finalize list of selected bidders, or select bidders from prequalification		
2. TECHNICAL SPECIFICATION • Write engineering specifications • Seek engineering specification approvals • Complete equipment data sheets • Complete drafting reference drawings • Assemble miscellaneous data • Complete technical requirements form • Obtain technical specification approval for printing		
3. COMMERCIAL DOCUMENTS • Compile commercial and financial sections • Complete contract conditions • Finalize contract program • Obtain commercial documents approval for printing		
4. BIDDING REQUIREMENTS DOCUMENTS • Write the invitation to bid • Write the instructions to bidders • Write the content of the bid • Check approvals for bidding requirements documents prior to printing • Assemble bidding document packages		
5. BIDDING PERIOD • Issue bidding documents for requesting bids		
6. BID EVALUATION AND DECISION • Evaluate bids • Negotiate with selected bidders as required • Draft and submit bid evaluation report • Authorized person to confirm successful contractor		
7. FINALIZE CONTRACT DOCUMENTS • Update all contract documents for issue		
8. AWARD CONTRACT • Write notice of award or letter of acceptance • Write letters of rejection to unsuccessful bidders		
9. CONTRACT AGREEMENT • Prepare contract agreement for execution		

FIGURE 4–1   Example of a bidding and contract document formation form.

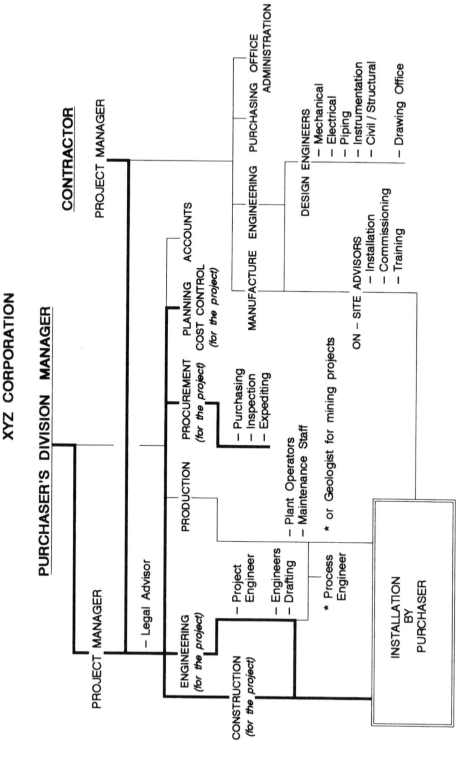

FIGURE 4–2 Example of purchaser and contractor reporting relationships for a major equipment procurement.

XYZ CORPORATION

IN–HOUSE DOCUMENT DISTRIBUTION FORM

PROJECT _____   **CONTRACT No** _____

In–House Procedures

Design Criteria

Technical Requirements

Contract Conditions

Project Program

Bidding Documents

Supplemental Notices

Written Confirmations

Contract Documents

Bid Evaluation Report

Bids Received

Record File

LEGEND IN BOX:

Original   [O]

Number of copies   [X]

Management										
Project Manager										
Project Engineer										
Design Engineers										
Procurement										
Planning										
Cost Control										
Construction										
Production										
Accounts										
Bidders										
Successful Bidder										
Unsuccessful Bidders										

FIGURE 4–3   Example of an in-house document distribution form.

COMPILING CONTRACT CONDITIONS
FOR PROCUREMENTS

5.0 INTRODUCTION

Based on the overall view presented in Chapter 1 concerning the formation principles of a legal contract, it should be recognized that no contract agreement should be finalized between the purchaser and a contractor for a major equipment procurement without the existence of a clearly formulated set of written rules that define the manner in which the contract is to be executed. These rules generally called *Conditions of the Procurement Contract* or simply, *Contract Conditions*, establish the rights, obligations, and responsibilities of both the purchaser and the contractor to a procurement contract. The contract conditions also define the course of action to be taken if either party fails to perform in accordance with the rules. It is extremely important that both the purchaser and the contractor be fully conversant with the intent of all the terms in the procurement conditions, and know how they will affect the procurement. Violations of the contract conditions by either party can lead to penalties for their organization. Ignorance of the regulations and law will not provide relief from damages that might be experienced.

With the continuous development of advanced technologies, together with large investments being made by industry and governments for the procurement of highly capitalized equipment, it becomes evident that the practice of issuing purchase orders with simple forms of standard in-house terms and conditions may not always be adequate for major equipment procurements. It is a matter of some concern how frequently matters of commercial or contractual importance have been ignored or overlooked.

There are a number of well-established standard model forms of general contract conditions that have been published for mechanical, electrical, and process engineering contracts, some of which are recognized internationally. These documents have been drawn up by representatives of all the interested parties concerned under the sponsorship of high-standing professional organizations. They are therefore accepted by owners, purchasers, and in particular, contractors and equipment suppliers. The United States does not have a nationally recognized standard set of general contract conditions for mechanical and electrical contracts similar to the ones mentioned above or those that have been prepared and published by the Engineers Joint Contract Documents Committee (EJCDC) for civil engineering and construction contracts. This may be due to the fact that no sponsoring organization has formulated a series of contract forms.

Much can be said in favor of the use of model forms of contract conditions as shown in later sections of this chapter. In the absence of a standardized set of general contract conditions that are applicable to most equipment procurements, it remains for each purchaser to draw up his own in-house set of "boilerplate" contractual rules. Since legal consequences are involved, the document must be written, or at very least, be carefully vetted by an attorney. The question arises as to the extent of provisions that are to be included in the document. Some believe that all relevant provisions should be noted, without defining "relevant," while others argue that this becomes a waste of time and paper. Their reason is that the law (the Uniform Commercial Code in the United States, for example) provides adequate remedies for most, if not all, practical occurrences. This may be true, but the contract conditions go further than just legal principles such as default, fraud, and nonperformance. They also cover procedural matters that are essential for the successful execution of the contract. The alternative is to run the contract without such procedures and risk making contract arrangements that are difficult to interpret and apply. This results in disputes because the trust that needs to exist between the parties to any contract is missing. The model forms of contract conditions mentioned previously contain important provisions to standardize both technical and procedural matters such as completion of work, taking over (from contractor to purchaser), acceptance, warranty, and so on, all of which are essential to major equipment procurements as further explained in Paragraph 5.1.1. The widespread use of standard contract conditions will not only facilitate the successful completion of a contract, but will in all probability result in lower bid prices being submitted as bidders become familiar with the conditions that apply without having to perform a detailed review of each purchaser's boilerplate conditions on which to bid and execute the contract. Standard contract conditions that have been published in a number of countries have been proved by experience to be satisfactory and are continuously refined by use over the years. The main point to remember is that these standard conditions are intended to apply to most contracts for equipment procurements, and that the scope of provisions do not overstep a purchaser's particular requirements for a contract undertaken.

It must be clearly stressed that procedures for the preparation of contract conditions and commentaries outlined in this chapter are intended to present general guidelines for a better understanding of the commercial aspects of engineering contracts. A more detailed and authoritative interpretation of key contract conditions can be found in books published on subjects relating to the law of contracts. However, the person responsible for compiling the procurement documentation should have acquired a knowledge of basic legal principles in the area of contract formation. He or she should be aware of the legal problems that are likely to arise during the currency of a contract and ensure that the documents are legally and contractually correct so as to avoid liabilities resulting from a breach of contract. The assistance and guidance of a competent legal advisor is always needed in the preparation and vetting of these documents.

When discussing the subject of contract conditions in general, there are a number of important issues relating to the financial aspects of a procurement. Because of their importance in the contract, they are dealt with separately in Chapter 6.

5.1 STRUCTURE OF THE CONTRACT CONDITIONS

Either the purchaser or the bidder or both can put forward contract conditions that each proposes to apply to the contract. Bidders very often prefer to qualify their bid with a set of their in-house terms and conditions, but it is usual practice for the purchaser to state in the bidding documents which contract conditions will govern the contract. These may be the purchaser's own contract conditions, or a standard model form of contract conditions that has been published by a national engineering institution or internationally recognized organization. While bidders can object to any provision clause and propose amendments and/or additions, such matters can be subject to negotiation with selected bidders after the completion of the bid evaluation, to arrive at a solution acceptable to both parties. Failing to reach a complete acceptance by both parties of the provision clauses will very likely lead to disputes during the course of the contract. Furthermore, once a contract has been established, the contract conditions cannot be unilaterally altered. Neither the purchaser nor the contractor may introduce new or amended contract condition provisions at a later stage without the agreement of both parties.

The contract conditions discussed are structured in three separate parts as follows:

1. *General (Standard) Conditions.* Provisions considered to apply to most equipment procurements undertaken and which must not be altered.
2. *Supplementary Conditions.* Deal solely with provisions for a particular procurement contract. Supplementary conditions also amend any general (standard) condition that does not meet the specific requirements of a particular procurement contract.
3. *Special Conditions.* Apply mainly to particular conditions such as company, statutory, or industrial policies that the purchaser has incorporated in a procurement contract. It is frequently mandated by law and regulations in some countries that the full text of an applicable statutory provision or regulation be included in the contract documents. The purchaser may also wish to do this voluntarily. Such statutory provisions can be added as an attachment to the supplementary conditions. As they concern matters that are of significant importance on their own, they are treated separately in this book as Special Conditions.

Many of the larger industrial organizations, as well as government agencies, have produced in-house standard or master conditions specially tailored to their operations and will have integrated the above mentioned contract conditions into the one document. With the use of word process software, a copy of the master conditions is amended to meet the requirements of a project and become the contract conditions for the particular contract undertaken. However, for the purpose of showing how contract conditions are developed for a major procurement contract, each of the above are discussed separately.

Hence, it is important to distinguish between the contract conditions that define how the work is to be executed and the technical scope of work to be performed by the contractor. One must have a clear understanding how major equipment procurements are engineered as compared with a simple "off-the-shelf" equipment procurement. It will then be appreciated why there is a need to keep technical and nontechnical documents separate. Unfortunately, experience has shown that engineers are primarily interested in engineering documents that directly concern them.

5.1.1   General (Standard) Conditions

There are a number of contract conditions that are applicable to most major equipment procurement contracts. These are commonly known as *General (Standard) Conditions.* While the wording and general intent of the provisions may vary, the objectives of these contract conditions should remain the same.

In the United States there are no national model forms of general contract conditions for equipment procurements other than those published by the Engineers Joint Contracts Documents Committee that are intended to be used for civil engineering and building projects. Some government agencies have contract condition documents incorporating standard provisions set out on a score of pages. To these are added a number of additional provisions to meet the needs of a particular contract.

Outside the United States, there are numerous model forms of general contract conditions in existence that apply to non–civil engineering, building, and construction contracts that have been prepared and published individually or jointly by national and broadly based internationally recognized organizations. They include:

- National and international engineering institutions and associations
- National standards associations
- National technical industry associations
- United Nations Economic Commission for Europe
- Organisme de Liaison des Industries Metalliques Europeenes (ORGALIME)
- Federal and state government agencies
- Large industrial companies

Sample general contract conditions have also been published by The International Bank for Reconstruction and Development (The World Bank) and the Asian Development Bank for contracts funded by the banks.

It is not the intent of this book to discuss the contents, distinctions, and merits of these standard documents, since each reflects the specific requirements of the individual organization that has prepared them. However, there are some standard model forms of published general contract conditions for plant and equipment procurements that have a narrow application with regard to methods of project operation and contractual requirements. As a result, substantial amendments may have to be made to these contract conditions to meet the requirements for a particular contract.

There are advantages for using standard model forms of general contract conditions. For the purchaser, the advantage is not having to establish and draft a new set of conditions for each and every contract undertaken. In the case of the contractors and equipment suppliers, it avoids the lengthy and time-consuming effort of having to review a new set of provisions each time when bidding for a contract. Furthermore, they become fully conversant with these documents because these contract conditions have been in use over a number of years and have been refined continuously by legal and technical experts to remove ambiguities and keep abreast of the latest technological trends so as to the meet the intentions of the contracting parties.

As a guide and possible checklist, the general (standard) conditions for major equipment procurements would include provisions to cover the following matters:

- Definition, and interpretation of terms used in the contract conditions
- Evidence of a contract—the contract agreement
- Compliance with country federal, state, or provincial laws governing contracts
- Ruling language
- Rights, duties, and responsibilities of the purchaser
- Obligations of the contractor
- Compliance with codes, standards, laws, and regulations
- Patents, copyright, and other intellectual property rights
- Confidentiality and advertising
- Assignment and subcontracting
- Use of contract documents:
  - Purchaser's documents
  - Contractor's documents
- Errors and omissions in documents supplied by the purchaser
- Contract variations
- Delays and extension of time
- Shop inspection, tests, and rejection—in general terms
- Passing of ownership of property
- Accidents and damage
- Completion of tests, taking over, and rejection of equipment

- Equipment warranty or defects liability—in general terms
- Certificates:
  - To record dates of contractual significance
  - For authorizing payments
- Limitation of contractor's liability
- Security for performance
- Force majeure
- Insurance—in general terms
- Suspension of contract by purchaser
- Termination for:
  - Default by the purchaser, or by the contractor
  - Insolvency
  - Convenience
- Resolution of disputes and arbitration

Explanatory commentaries on selected major condition provisions are discussed in Paragraph 5.2.2.

5.1.2   Supplementary Conditions

Conditions relating to a particular contract in question, such as the description of the project, contract price structure, time for completion, delivery, and the like, are drafted individually by the purchaser and are preferably contained in a separate document titled *Supplementary Conditions*. These conditions concern those terms and conditions that are to be applied to the contract over and above the general (standard) conditions. They also include certain rules that the purchaser requires to be applied to a particular contract covering such matters as administration and control procedures.

Drafting supplementary conditions requires a complete understanding of the general conditions in order to determine the supplementary conditions that are required for the project. Whereas some of the provision clauses in the general conditions are presented in a standard form, such as insurance, specific requirements for insurance for the project are dealt with in the supplementary conditions. In cases where a provision is covered both in the general and supplementary conditions, care should be taken to ensure that no ambiguity or conflict arises between the provision clauses. It should be clearly stated, preferably in the contract agreement, that where a supplementary condition is not consistent with a general condition, the supplementary condition takes precedence. Furthermore, terminology must be consistent in both the general and supplementary conditions or the latter provisions may be misinterpreted or not be fully understood as to intent. It is therefore recommended that legal assistance be sought on the precise wording of the supplementary provision clauses prior to their release for bidding purposes.

It should be noted that some engineering institutions and industrial organizations use the term "Special Conditions of the Contract" to supplement the general (standard) conditions. The use of "special" is discouraged as the primary purpose of these conditions is to supplement the general conditions for a particular contract. Furthermore, they may be in conflict with the special conditions described in Paragraph 5.1.3.

Supplementary conditions that apply to a particular equipment procurement would normally include:

- Contract price structure
- Contract price adjustments for cost variations
- Terms of payment
- Contract program
- Liquidated damages and bonus payment
- Particular indemnification
- Particular insurance requirements

Work-related procedure provisions may include:

- Progress control and reporting
- Application for payment
- Change order procedures
- Claims procedures

These provision headings can be set in the form of a standard schedule since they can apply to most major equipment procurements undertaken. The schedule itself becomes an integral part of the supplementary conditions. Certain fill-in spaces are completed by the purchaser before the bidding documents are released for soliciting bids. It is finally completed prior to award of contract with the information provided by the successful bidder.

The schedule summarizes the key contractual data for the particular contract and is identified as the Attachment to the Supplementary Conditions. It will also serve the purpose of a checklist of items prior to requesting bids or items to be negotiated for each contract. The attachment must be subject to agreement by both the purchaser and the bidder before issuing the purchaser's notice of acceptance of the bid.

The suggested fill-in spaces on the attachment sheet cover such items as:

- Contract number
- Name of the purchaser
- Notices—address of the purchaser
- Name of the contractor

- Notices—address of the contractor
- Applicable law (country/state/province)
- Language for day-to-day communications
- Time for delivery
- Place for delivery
- Time within which the purchaser must give a decision for contractor's drawing and technical data submittals
- Time for payment of claims
- Contractor shall provide security in the amount of (state amount)
- Contractor's limits of liability
- Liquidated damages for delay
- Liquidated damages for not achieving guaranteed performance
- Limit of liquidated damages
- Retention moneys
- Insurance covering equipment
- Insurance covering public liability
- Person or organization nominated as arbitrator
- Procedural law for arbitration
- Place for arbitration

Note: Contract price and other pricing data are normally noted in the purchaser's notification of award of contract or letter of acceptance.

5.1.3   Special Conditions

Special Conditions refer to company policies the purchaser requires the contractor to adhere to, or statutory regulations that may apply to special circumstances associated with the contract. While the contractor is required to adhere to all statutory laws, regulations, and requirements that prevail for the work under the contract, there may be certain provisions that should be drawn to the contractor's attention.

Examples of special conditions may include:

- Racial discrimination policy
- Utilization of prison labor
- Purchasing policy
- Environmental, pollution, and protection regulations

- Occupational safety and health regulations
- Nuclear regulations

5.2   EXPLANATORY COMMENTS ON KEY WORDS AND PROVISIONS

5.2.1   Definition of Key Words and Interpretations

Contract conditions documents should begin with a list of definitions of key terms used in the documents. This is necessary because there is no known standardization of key contract terms used in contract conditions documents, and can result in one purchaser's interpretation of terms being substantially different from another's. It becomes extremely important, therefore, that each set of contract conditions be provided with definitions of the key terms used, to ensure that the same meaning is assigned to them and to save any legal challenges as to their interpretation.

The following is a list of selected key words and expressions commonly used in procurement contract documents. Many others are self-explanatory and are not mentioned herein. It should be noted that these key words in contract documents are intended to have a special meaning and are therefore shown with uppercase initial letters. When intended to have their normal dictionary meanings, they are shown with lowercase initial letters.

- *Contract* shall mean the agreement between the purchaser and contractor for the execution of the work incorporating all the contract documents physically attached thereto.
- *Contract Agreement* shall mean the document evidencing the terms of the contract between the purchaser and the contractor and signed by the legally authorized persons.
- *Contract Documents* shall mean all the documents referred to in the contract agreement.
- *Contract Price* shall mean the sum named in the notification of award of contract, or letter of acceptance of the bid.
- *Contract Value* shall mean such part of the contract price adjusted to include priced additions or deductions as made during the course of the contract.
- *Contract Program* shall mean a contract document recording milestone dates which the contractor as well as the purchaser have bound themselves to meet. It will record the start and completion dates of the contract, and for any major section or sections of the work that have to be ready by the specified times or dates noted in the document.

- *Purchaser* shall mean the person or party named as such in the contract agreement requiring the work to be executed.
- *Contractor* shall mean the person or party named as such in the contract agreement who will carry out the work for the purchaser.
- *Subcontractor* shall mean any person or party named in the contract as a subcontractor, or with the written consent of the purchaser, engaged by the contractor to execute part of the work under the contract on behalf of the contractor.
- *Procurement* shall mean the series of activities for the purchase of equipment from the commencement of preparing the bidding documents through to delivery of the equipment at the purchaser's nominated site or sites.
- *Work* shall mean the whole of the work including the furnishing of and delivering of the equipment, and services, to be executed by the contractor in compliance with the contract.
- *Site or Project Site* shall mean the geographical location where the equipment is to be delivered in compliance with the contract documents.
- *Specification* shall mean the specification for work to be executed by the contractor under the contract as existing at the Date of Acceptance of the Bid and any amendment to the specification made thereafter by the purchaser.
- *Drawings* shall mean drawings supplied by:
    - (a) The purchaser to the contractor forming part of the technical specification for the purposes of the contract
    - (b) The contractor to the purchaser under the terms of the contract
- *Documents* shall mean all printed material other than drawings.
- *Documentation* shall mean any relevant documents, and drawings depending on the circumstance, including the preparation of documents for bidding and implementation purposes, and where appropriate, technical documents such as engineering specifications, calculations, drawings, equipment data sheets, reports, or documents produced by data systems.
- *Bid* shall mean the priced offer submitted by the contractor to execute the work in accordance with the bidding documents.
- *Date of Acceptance of Bid* shall mean the date that appears on the purchaser's written notification to the contractor of acceptance of the bid.
- *Date/Period for Delivery* shall mean where the contract provides a date for delivery to or from a point of destination, the date or the extended date in accordance with the provision of the contract, *or*, where the contract provides a period for delivery, the last day of the period for all the deliveries or for last of the work components.
- *Ex Works,* as used in this book is a term in Incoterms,[1] means that the contractor

---

[1]Ex Works, FOB, CIF, and other trade terms are defined in Incoterms detailed in Appendix C.

fulfills his obligation when the equipment has been made available at his or the manufacturer's premises (works, factory, warehouse, or elsewhere) to the purchaser. He is not responsible for loading the equipment or any part thereof on the vehicle provided by the purchaser unless specified otherwise in the contract documents.

- *Taking Over* shall mean the transfer of ownership risk, and responsibility for the equipment from the contractor to the purchaser when all tests have been passed and all work under the contract is complete in all material respects and is accepted by the purchaser. The equipment warranty or defects liability period starts from the date of taking over. From that time, the purchaser is responsible for the care, operation, maintenance, and safety of the equipment.

- *Equipment Warranty Period* shall mean the period during which the contractor is responsible for rectifying any defect of the equipment in accordance with contract conditions and the technical specification as defined in the contract.

- *Risk Transfer Date* shall mean the date when the risk of loss of, or damage to, the work or any section thereof passes from the contractor to the purchaser.

- *Written or in Writing* shall mean any communication or statement handwritten, typewritten, printed, or by electronic telegraphic means.

- *Day* shall mean a 24-hour calendar day commencing at 12:00 midnight.

  Note: An extension of time for the completion of the work may need to be made in terms of working days which normally exclude Saturdays, Sundays, and proclaimed holidays.

- *Week* shall mean any period of seven consecutive calendar days.

- *Year* shall mean any period of 365 (or 366 for a leap year) consecutive calendar days.

Reference Dates:

- *Contract Date* (Day 1 of the contract) is the date of the purchaser's written notification to the successful bidder stating acceptance of the bid, unless specified otherwise in the contract documents.

- *Base Date* is the date stated in the supplementary conditions for calculating the CPA (contract price adjustment) in accordance with labor/material cost formulae noted in the supplementary conditions.

5.2.2   Explanatory Comments on Selected Contractual Provisions

This section contains a brief description of selected provisions common in most major equipment procurement contracts, and which have not always been fully understood as to their meaning and intent, or how they are to be applied. It must be emphasized that the information and analysis are meant for guidance purposes only. The aim is to assist both the purchaser's engineer and the contractor to attain a better understanding and application of the principles underlying these contract provisions.

It is not the intention for these comments to replace or modify contract conditions that have been published by private or government organizations. They have legal implications and those responsible for preparing this work should always seek the help and guidance of competent legal advisors.

The following are discussed in alphabetical order:

- Arbitration
- Assignment of the contract
- Completion, taking over, acceptance, and certificates
- Contract documents
- Documents—supplied by the purchaser and by the contractor
- Force majeure
- Inspection and tests
- Liquidated damages
- Subcontracting
- Variations to the contract
- Warranties (or guarantees)

A general comment is made also on specifying times for submission of documents.

Arbitration

Disputes arising between the purchaser and the contractor under the contract that cannot be settled should be resolved, if possible, without resort to litigation. To accomplish this, the contract conditions should contain a provision setting out arbitration rules and procedures to be followed by the parties should such situation arise.

Arbitration means the submission of a dispute to a person or persons who have no financial interest in the final decision of the dispute. This entity, called the Arbitrator, shall have full authority to review decisions, opinions, and directives. The decision of the arbitrator, where acted on within the authority prescribed in the contract conditions, should be taken as conclusive and final, though, should one party fail to accept the arbitrator's decision, the opportunity still exists for the claim to become one for the courts to settle. It should be noted that arbitration is generally favored by most courts and an arbitrator's decision would only be overturned in extreme and exceptional circumstances.

The name of the nominated arbitrator or arbitration association should always be stated in the contract conditions and be acceptable to both parties. Should arbitration be in accordance with an arbitration authority such as the International Chamber of Commerce, the title of rules under which the arbitration is to be carried out should be stated.

Arbitration may be resorted to even while work under the contract is progressing. Indeed, the contractor should continue with the work and likewise, the purchaser continue to make payments to the contractor as and when they become due under the terms of the contract.

Matters that in general are cause for dispute include:

- What is included and/or not included in the contract
- Interpretation of specifications, drawings, and other documentation forming part of the contract
- Claims for variations—extras and deletions
- Claims for delays
- Payment claims at specified contract completion stages
- Claims for liquidated damages

An issue considered sufficiently important for inclusion in the contract conditions is timing of arbitration submissions. Where there is a dispute, the provision should make reference that the contractor should (or must) give 14 days notice before lodging a claim under arbitration if the purchaser fails to respond to the contractor's claim for settlement. Should no times be specified, there can be claims coming from anywhere at any time and for little reason. The time limit restricts such occurrences in general and allows for reasonable actions to be taken by both parties.

In summary, arbitration becomes a fast, business-like, and inexpensive means for settling disputes.

Assignment of the Contract

This is a prohibition against the contractor passing on his contractual obligation to some other contractor to perform all or any part of the works under the contract by assignment of the contract without the prior written approval of the purchaser. As the purchaser selected and awarded the contract to the contractor following successful negotiations that led to the signing of the contract agreement then accordingly, the contractor is committed and responsible to perform all parts of the work.

It is not unknown that after acceptance of the contract, the contractor's workload is such that he has to seek someone who will assist in meeting his obligations under the contract. Notwithstanding this occurrence, approval must always be sought from the purchaser.

While the assignment prohibition applies to the contractor, of equal importance is that the provision also implies that the purchaser cannot assign any obligation to another party without the contractor's agreement.

Completion, Taking Over, Acceptance, and Certificates

Considerable confusion and misunderstanding exists with the meaning and practical application of the term "completion" of work by the contractor, "taking over," and "acceptance" of the equipment by the purchaser. The diverse use of these terms appears to be a result of lack of agreement being made to internationally standardize the definition and interpretation of these important milestone events in the contract.

The concept of a *Contract Completion Date* of the work by the contractor can be considered to be achieved when the following key sections of the contract are finalized:

- That nothing has been omitted from what has been specified in the technical specification
- That the work has been finished within the time specified in the contract program
- When the equipment has successfully passed specified performance tests
- When the purchaser commences operating the equipment and the contractor is no longer responsible or liable for ongoing operation and normal maintenance of the equipment, *or*
- The contract is considered complete after the expiry date of the equipment warranty or liability period, all defects having been corrected

As each of the above is a major event in the contract program, it is of utmost importance that there are provisions in the contract conditions that clearly define what constitutes completion, and the procedures for administering and officially documenting the events. In general terms, completion is usually affected when the contractor has fulfilled all contractual obligations within the *Time for Completion,* and this forms the basis for any claim the purchaser may make for delay in completion of the work, and for any extension of time claimed by the contractor.

One of the most significant events in the contract program is the transfer of ownership from the contractor to the purchaser, and is termed in this book as *Taking Over* the equipment. The term "practical completion" is also commonly used as well as "mechanical completion" in this context. The latter should be avoided as taking over normally comprises all of the equipment of which mechanical equipment may only be a part. The purchaser's precise procedures and documentation to evidence taking over the equipment must also be defined in the contract conditions.

It is the contractor who usually serves notice to the purchaser that in his opinion, the equipment is complete for taking over. He does this by making an application for the purchaser to issue a formal *Taking Over Certificate*. The date of the taking over must be noted in the certificate, and not the date of the certificate, as it is at the date of the taking over that the purchaser becomes the owner of the equipment and is now responsible for its care, operation, normal maintenance, and safety. It is also the date that the equipment warranty or defects liability starts and the contractor's obligation apply to rectify any defect in the equipment as defined in the contract conditions.

For an equipment supply and deliver contract, and the equipment having been performance tested in the manufacturer's shop, the diagrammatical structure of taking over and equipment warranty or defect liability is shown in Figure 5–1.

Even though the equipment is complete at the taking over by the purchaser so that it can be used for the intended service, there may well be defects or uncompleted items of work that are of minor substance and that the contractor is still required to correct or complete. These items are to be listed on the taking over certificate with an instruction to the contractor that they must be corrected or completed within .... days. If the contractor fails to do so within the specified period, or within a reasonable time, the purchaser can arrange for the outstanding work to be carried out utilizing his own organization or by utilizing others, and deducting the cost of the work from the contract price.

The term *Acceptance* of the equipment is used in this book after the successful comple-

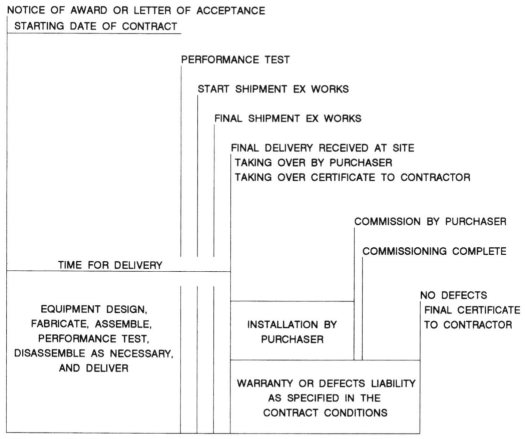

FIGURE 5–1 Diagrammatic phase structure of contract milestones for a major plant equipment supply and delivery, with performance tests performed in the manufacturer's shop.

tion of performance tests at the project site. To evidence this, the purchaser issues an *Acceptance Certificate* as shown in Figure 5–2.

Should the equipment fail to achieve the contractor's guaranteed performance and after repeating the tests, the purchaser may be prepared to accept liquidated damages as a remedy if the equipment falls up to .... percent below guarantee (see Chapter 6, page 6–13) and then issue the acceptance certificate. The purchaser should be free to reject the equipment if there is a greater loss than that stated in the contract. If the purchaser requires this

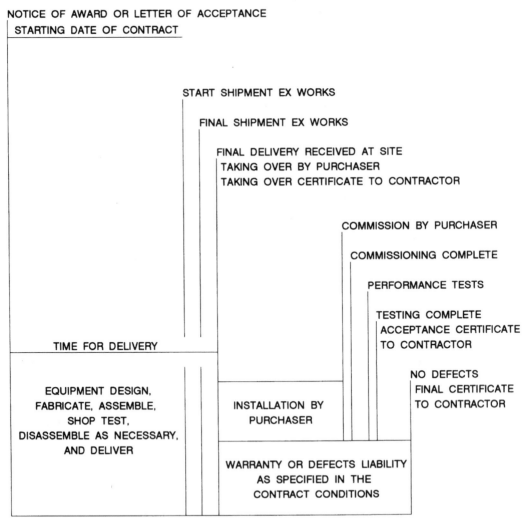

FIGURE 5–2 Diagrammatic phase structure of contract milestones for a major plant equipment supply and delivery, with performance tests performed after installation.

right, then it should be expressly stated in the contract conditions making also reference that this express right of rejection is subject to the maximum sum of liquidated damages having been exceeded.

As soon as the equipment warranty or defects liability has expired, the contractor serves notice to the purchaser that he has made good all defects, and requests the purchaser to issue a certificate called the *Final Certificate*. This certificate states the equipment is finally complete as from the date thereof. Should any part of the equipment have been subject to repair or replacement by the contractor during warranty period, the contract conditions should clearly state that the affected part has to run afresh for a defined warranty period as from the time rectification was completed. The purchaser should nevertheless issue a final certificate for the remainder of the equipment.

Finally, the contractor makes an application for a *Final Certificate for Payment*. In some contracts, the final certificate is also a claim for final payment. All adjustments to the contract price must be included in the final certificate for payment. The intent of issuing the final certificate for payment is to evidence as final, the proper fulfillment by both the purchaser and contractor of their obligations under the contract.

It is normal for the contract conditions to state that no further claims will be considered by the purchaser after the final certificate for payment is made.

In either case, taking over the equipment can only occur after the work has been completed and accepted by the purchaser. It is at this moment that the risk of loss of or damage to the equipment passes to the purchaser, and the contractor's obligation in relation to the equipment warranty or defects liability period becomes effective (see Warranties, page 5–23).

The contract conditions usually have provisions for certificates to be issued by the purchaser to the contractor to record the date when a contractual event of significance occurred, such as the taking over and to make any payment the contractor may be entitled to at that point. The contract conditions should state the number of days within which the contractor must apply for the certificate, and for the purchaser, the number of days after receipt of the application to issue the certificate.

When the equipment has been taken over, the purchaser issues a *Taking Over Certificate* with the date shown certifying when the equipment was taken over. It is from that date that the contractor's obligations in relation to the warranty or defects liability will commence. Even though the purchaser has issued a taking over certificate, there may well be defects of a minor character or items remaining outstanding that the contractor is required to correct or complete. These items should be listed in an attachment to the certificate with an instruction to the contractor to complete or correct them within .... days.

The purchaser issues a final certificate to the contractor at the end of the equipment warranty or defects liability period to record the end of the contractor's obligations under the contract.

Contract Documents

Chapter 3 stated that a contract should always list the documents that define and form the contract. These documents were regarded as the contract documents, and those documents considered in this book were shown in Figure 3–2.

The need to exercise extreme care in the preparation of the contract documents is obvious since correct interpretation of the contents of each document applies not only to the contractor but also to the purchaser's staff assigned to the contract. While the person writing the particular provision requirements knows what each is intended to mean, they can be interpreted in an entirely different way by others. Chapter 14 identifies and lists key points that will be helpful to one who is writing the documents.

Drafting the bidding and contract documents normally involves a number of persons contributing to different sections of the work. However, it remains the responsibility of the purchaser's engineer forming the contract for the correctness of the documents and to ensure that there is no ambiguity, discrepancy, or conflict of information that may arise between the documents. Should the contractor discover any ambiguity or discrepancy, the matter should be brought immediately to the purchaser's attention. The purchaser should then clarify the issue and supply the appropriate interpretation to be followed by the contractor. It is assumed that such a directive may never be cause for the contractor to incur a greater expenditure than the successful bidder could have reasonably anticipated at the time of preparing the bid, but if there is any additional expenditure then it may be subject to a claim for an addition to the contract price.

While the contract documents in a contract complement one another, the matter of contract document precedence is one of major importance and must not be overlooked when compiling the contract. Some standard model forms of general contract conditions not only list the contract documents but also specify the order of precedence or priority of the documents. It is often shown that the supplementary contract conditions take precedence over the technical specification. The issue is debatable. Some lawyers may form the opinion that legal issues are more important than the technical requirements. Engineers who draft the technical specification for a particular project will consider their work to take precedence over legal or commercial matters. However, for a procurement contract where the contractor is responsible for producing a design that conforms to the purchaser's technical requirements, it is considered that the main emphasis of the contract is on technical issues in preference to commercial matters.

Should the purchaser wish to alter the order of precedence from the one noted in the general (standard) conditions, the relevant clause(s) will have to be amended in accordance with the suggested procedure described in Paragraph 5.1.2. It would be more logical to incorporate the precedence of documents in the contract agreement.

Chapter 12 makes specific reference to revising both purchaser's documents and the successful bidder's bid where required to incorporate items that were subject to negotiation prior to the purchaser and the contractor signing the contract agreement. The importance of accurately performing this task cannot be overemphasized.

Documents—Supplied by the Purchaser and by the Contractor

Unfortunately, many contract conditions make reference to the term "drawings" to refer to all types of documents supplied by the purchaser (other than contract documents) and those to be supplied by the contractor. To prevent this misinterpretation, it is recommended that the terms "drawings" and "documents" be defined in the contract conditions document as noted in Paragraph 5.2.1.

It is important to recognize the clear distinction between contract conditions that concern both the purchaser and the contractor on matters relating to technical documentation, and the contractor's obligation to supply this information as part of the scope of work set out in the technical specification. Some contract conditions have provisions detailing contractor's scope of work requirements with respect to drawing and data submittal requirements, and this has raised the question: Where do the contract conditions end and where does the technical specification start? This is a problem area that is further discussed in Section 5.3.

It is suggested that the contract conditions make reference to such matters as:

- Confidentiality, ownership of proprietary information, patent rights, and the preparation and transfer of "as built" drawings at or on the completion of the contract and their use thereafter. Should the purchaser seek ownership of the contractor's information, this may involve a separate arrangement possibly leading to a licensing agreement.
- Media release of documentation or oral communication
- Mistakes and discrepancies in documents
- The period of time (usually expressed in days) for the purchaser's review from the date the information is to be received by the purchaser. It is the contractor's responsibility to have the information available by the specified date regardless of the method of transmission (post or courier). It is the purchaser's responsibility that the reviewed information is returned and received by the contractor within the specified time and to advise the contractor of the review status. If not received by the contractor within the specified time, the contractor will be entitled to assume that the information is deemed to be accepted without comment. However, this should always be stated in the contract documents.

Exceeding the specified period for receipt of the purchaser's review of the contractor's documents may not always constitute an automatic approval by the purchaser. Automatic approval without confirmation can lead to an extremely complex situation. Commencing work prior to the purchaser's comments constitutes a contractor's risk, but any lag on the part of the purchaser can delay completion of the procurement. Such requirement is particularly important especially for the procurement of large and complex equipment. It may also lead to claims for extension of time to the contract completion date.

Force Majeure

Few appear to be familiar with the term *Force Majeure* and even fewer know how to apply it in practice. The term force majeure covers unforeseen circumstances or any circumstance beyond the control of the parties arising during the course of a project that result in either one or both parties being partially or totally unable to fulfill their obligations under the contract. Such circumstances include, but are not limited to: breakdown of plant, act of God, fire, explosion, war, rebellions, riots and other uprisings and disorders, and industrial action by workers or employers. It should be noted that a mere shortage of labor or materials does not constitute a force majeure unless it is caused by circumstances which are themselves force majeure.

Where the contractor is unable to meet obligations under the contract, he shall be entitled to give the purchaser a *Force Majeure Notice* within 14 days of the occurrence, setting out as fully as possible the circumstances alleged to constitute the force majeure. Should the purchaser dispute the contractor's right of the force majeure notice, he shall give the contractor notice to that effect within 14 days after receipt of the contractor's notice. Should the contractor contest the purchaser's reason, the dispute shall be referred to arbitration in accordance with the contract conditions.

The contract conditions should make reference to the settlement of possible additional costs incurred by the contractor as a result of the force majeure, and that the contractor shall not be liable for any damage or loss suffered by the purchaser arising from the reasons stated on the force majeure notice. Further, the possible claim for extension of time for completion of the work should be addressed. Contract clauses should also define possible contract termination provisions in consequence of a force majeure, such as payments to the contractor and release from performance bonds. Certain countries have laws stipulating the consequences of a force majeure.

Inspection and Tests

It must be recognized that the purchaser or his representative has the right to have unobstructed access to the manufacturer's works for the purpose of inspecting materials, workmanship, and other items pertaining to the contract to confirm their conformity with the engineering specification. This right is often noted in the contract conditions. Not to be included in the contract conditions, however, is the scope of inspection and testing requirements set by the purchaser. These should be clearly specified in the engineering specification and/or equipment data sheets.

The purchaser's inspector or his representative has the right to reject any part of the work that fails to conform with the engineering specification and to have replaced or have made any alteration necessary to conform to the engineering specification, without cost to the purchaser.

Should an inspector reject any part of the work, a written statement should be forwarded to the contractor stating the reason for the rejection. However, an inspector's rejection is not necessarily binding or final and may be challenged by the contractor and become subject to arbitration.

Regardless of the release of any part or the whole of the work by the purchaser, the contractor shall not be relieved of any of his obligations under the contract.

Liquidated Damages

In many major procurements, the bidding documents state that "time and performance achievements are of the essence in this contract." The contract may then include provisions for the purchaser to be entitled to claim appropriate damages from the contractor if either or both of the following events arise from failure:

- To complete the work under the contract by an agreed date
- To achieve the agreed guaranteed performance of the equipment under the contract

The damages are the pre-estimated or predetermined amounts agreed to by both the successful bidder and the purchaser at the time of bidding, for the full amounts the purchaser might be expected to suffer by way of loss of income by reason of the contractor's failure to meet his contractual obligations set out in the contract. The amounts written into the contract before the contract agreement is signed are referred to as Liquidated Damages.

The agreement reached by the successful bidder and the purchaser on the amount of any damages in advance of placing the contract avoids the situation of having to settle a dispute on the matter by an arbitrator or a court, should such an event arise.

Rates of liquidated damages must be genuine amounts reached by agreement between the successful bidder and the purchaser. They can be less than the agreed amount, but never exceed the justifiable estimates. If the amounts are in excess of the greatest loss that could be proved, they cease to be a recovery of damages and become a penalty or punishment by the purchaser on the contractor. Such action may not be mandated under legal systems in some countries.

For most major contracts, the amount of liquidated damages to compensate the purchaser's potential financial losses are not related in any way to the value of contract. As for the successful bidder's acceptance of the maximum amount of liquidated damages payable to the purchaser, it must be appreciated that no contractor can be expected to insure himself for any unforeseeable risk during the course of the project, for amounts that may be completely out of proportion to the value of the contract or to his anticipated profit.

The purchaser will generally deduct any liquidated damages from the sum due to the contractor at the time of final payment though, in some cases it may be necessary to take these from progress payments.

It should be noted that certain legal systems have rulings relating to the extension of time for completion of the work, and the enforcement for payment of liquidated damages by the contractor. The issue can go further when the extent of a delay in completion becomes excessive and the purchaser considers this to be a breach of contract.

Methods of applying liquidated damages to a contract are dealt with in Chapter 6.

Subcontracting

Following a similar philosophy to the prohibition against assignment, the contractor should not be permitted to subcontract all or any part of the work without the prior written approval of the purchaser. Normally, the purchaser should not withhold approval from the contractor from subcontracting part(s) of the contract work, especially to those who specialize in that particular area of the work. On the other hand, the purchaser should not permit the contractor to unlimited subcontracting for work that the purchaser knows can be better performed by the contractor than by any subcontractor.

The bidder is required to list the items of the work to be subcontracted and the name of the subcontractor who will perform the work in a schedule to be included with the bid (see also Chapter 8). This information is analyzed during the bid evaluation process, especially the qualifications and financial position of the nominated subcontractor(s).

If the bidder is unable to gain acceptance of his nominated subcontractor(s) during the bid negotiating stage, an acceptable substitute may have to be furnished (see also Chapter 11). Hopefully an acceptable substitution will not be cause to alter the bid price, but for certain reasons the bidder may be entitled to an adjustment in price because of the change. In all circumstances, the purchaser can protect his interests, not only with regard to the subcontractor but also the contractor, by the issuance of a performance bond as further discussed in Chapter 13. No doubt, the surety company will make a thorough investigation of the contractor and the nominated subcontractor's technical ability as well as financial soundness before issuing a performance bond.

There may be instances where the bidder has not finalized the selection of a subcontractor by the time the bid has been submitted. The purchaser should still maintain the right to approve or reject a subcontractor nominated after award of contract.

Variations to the Contract

It becomes a great burden to the contractor if the purchaser changes the scope of work during the course of a lump-sum contract. It is obvious a contractor would prefer that the contract agreed to is final and will not vary.

There are very few problems, frustrations, and anxieties that have caused greater concern in the administration of a contract than variations. They are blamed for cost overruns, delayed completion of the contract, and disputes. The need to make contract alterations frequently stems from the purchaser's unseemly haste to "get the project started" without proper planning and/or lack of sufficient attention to the detail and accuracy of specifications.

Obviously, variations to the contract are very frustrating for the contractor when completed (or near completed) design work has to be aborted, orders cancelled or renegotiated, deliveries delayed, and the additional work costed and reprogrammed.

When changes to the contract are unavoidable, a change order issued by the purchaser should be executed on a standard form and addressed to the contractor, dated, clearly outlining the work to be modified, adjusting the contract price and time for contract completion if necessary, and be signed by the purchaser's authorized person. The change order should have a space for the signature by the contractor acknowledging receipt of, and agreeing to the terms of the change order.

Most contract conditions have a provision whereby the purchaser reserves the right to make changes in the scope of work and the contractor agrees to the purchaser's right to do so. This is not unilateral as a further contract condition provision should state that the contractor will be paid for making the changes that have been requested by the purchaser and for any additional work to the specified requirements. There should also be a provision stipulating that the purchaser receives a credit if the changes result in a lower contract price for the work. The net result is that the purchaser may order a change to the scope of work at any time and the contractor is obliged to proceed with it whether or not he agrees. The contract conditions should establish procedures as to how a change order will be processed so that it is acceptable to both parties.

Warranties (Guarantees)

The terms *Warranty* and *Guarantee* are used almost interchangeably in both contract conditions and technical specifications. When one looks up the definition of each in recognized dictionaries, the issue becomes even more confusing. As a result, many professional organizations have preferred to use the term warranty.

Special attention must be given to warranties in a contract not only because of their financial importance but also because of the laws regarding warranties, which vary by state in the United States. Hence, the formulation of any warranty provision must be in accordance with the proper state law. Of particular note is the legality of penalties without a corresponding bonus for better than specified performance. The purchaser should always secure an opinion from a competent lawyer regarding the wording of any warranty provision in the contract documents.

In general, there are three classes of warranties:

1. Delivery schedule warranty
2. Performance warranty
3. Equipment warranty, also known as defects liability

A contract may have a *Delivery Schedule Warranty* where a penalty/bonus provision for the delivery of the equipment, whole or parts thereof, is considered to be of major importance. This provision in the contract conditions covers a specified penalty for each day delay in a given event and a bonus for each day early. The schedule must be realistic with the date of the event defined and achievable by the contractor. Special allowances must be made for force majeure matters that the contractor cannot foresee.

A *Performance Warranty* is usually a provision that the contractor warrants that the equipment will be capable of achieving a required level of performance, and carries with it a financial liability should the equipment fail to perform as warranted. A bonus for exceeding it may be required by law. The warranted level of performance must be clearly specified in the technical specification being a function of the design of the equipment as well as detailing as and when the equipment is to be tested (for example, after a specified period of uninterrupted operation), the method and responsibilities for testing.

If the equipment fails any repeated test(s), it usually triggers the payment of liquidated damages set in the contract conditions. The liquidated damages provisions must state the maximum the purchaser is prepared to accept as a remedy for any equipment performance deficiency, and also in the event of any greater deficiency (a defined point), the purchaser's right to reject the equipment.

An *Equipment Warranty*, also referred to as *Defects Liability* or *Defects Correcting Period*, is a provision provided by the contractor, that the equipment will be free of fault for either:

- A specified period of time (calendar months or years), *or*
- A specified number of hours of operation (or equivalent), *or*
- A specified quantity of production or throughput quantity (e.g., tonnes production)

The definition of a fault extends to defects in design, material, and workmanship covering the whole area of the contractor's responsibility. It excludes mal-use, and misoperation of the equipment by the purchaser. The above specified conditions must be incorporated in the technical specification as each may be a criterion for design and/or the selection of a particular material. All too often, the contract conditions, standard or otherwise, commonly specify a 12 month period, which may be in disagreement with the technical specification and must be coordinated. It has become current practice for the purchaser to specify a period as long as can be negotiated with the contractor before award of contract since the purchaser may suffer appreciable financial losses while the equipment is being replaced or repaired.

Matters relating to the start of the equipment warranty, making good defects, extension of the warranty, delays in remedying defects, and limitation of liability for defects should be dealt with in the contract conditions.

General Comment on Specifying Times

Care should be taken when specifying a standard time period for issuing and submitting notices and documents. Notices can be sent by facsimile transmission, but document delivery must take into account distances between the purchaser's and the contractor's locations, especially for overseas contracts. Document transmittals can be delayed through strikes and other acts of God. Notices should always be sent advising the party that a document transmittal has been forwarded (advise date and the means of conveyance).

5.3   MAJOR PROBLEM AREAS BETWEEN CONTRACT CONDITIONS AND THE TECHNICAL SPECIFICATION

It must be recognized that certain contract conditions for equipment procurements will have provisions that make reference to related technical matters in order to fully define the condition and how it will affect both the purchaser and the contractor. However, utmost care must be exercised to distinguish whether the technical matter is affiliated with the contract rules or is a scope of work provision to be performed by the contractor forming part of the technical specification. This becomes an important issue especially for the contractor, as one must have a clear understanding of how and by whom the commercial and technical documents will be used after award of contract. For a major equipment procurement the contractor's design and engineering personnel will rarely be concerned with any documentation other than the technical specification. On the other hand, the contract conditions are more a matter for application by managers and for the overall administration of the contract.

An incorrect mix of commercial and technical provisions in the bidding documents can cause profound confusion to the end users. This is because applicable sections of the work are often removed from documents and distributed to those who are directly involved with a specific part of the work. Unless accurate distribution records have been maintained, it can become extremely difficult to determine who is to receive a purchaser's revised document. Such distribution would include not only the contractor's personnel and subcontractors (if any), but also the purchaser's expediters and inspectors assigned to the project.

Unless the complete set of bidding documents is compiled by the one person, it is important to ensure that the technical and nontechnical provisions be coordinated by those responsible for the input of their respective data. Whereas certain provision duplication may be unavoidable, under no circumstances must there be ambiguity or conflict of requirements. Unfortunately, this issue has not always been adequately addressed either through ignorance of procedures or due to haste to release the documents regardless of possible consequences.

The main problems are in:

- Contract administrative procedures
- General work related issues—scope of work requirements

5.3.1   Contract Administrative Procedures

Specific administrative procedures are developed by the purchaser for the administration of a particular contract. They normally include matters such as:

- Procedures for applications for payment
- Measurement, and evaluation of work for payment
- Contract variations and change orders
- Taking over, and certification
- Progress reporting—affecting the contract program but which can be duplicated with the technical specification as a scope of work requirement

These provisions can be included in either the supplementary conditions or contained in a separate document titled "Administrative Procedures." The latter would become an additional contract document and this is not recommended.

5.3.2   General Work-Related Provisions

Work-related provisions are those considered to be duties that the contractor is required to perform as part of the scope of the work under the contract. As they would have been cost estimated by the contractor for inclusion in the contract price, these provisions would then be incorporated in the technical specification to cover such matters as:

- Quality control inspection and shop tests
- Drawing and data submittal requirements

- Operating and maintenance manuals
- Crating and shipping
- Equipment performance testing—scope, procedure, and testing responsibilities
- Defects liability—period in months, *or*
  - minimum hours of operation, *or*
  - to achieve a minimum production, say in tonnes for material wear
- Spare parts

6

DEVELOPING COMMERCIAL TERMS RELATING TO PROCUREMENTS

6.0 INTRODUCTION

Of utmost importance to the success of any contract is its profitability. If the purchaser does not provide the required commercial information and requirements in the bidding documents, it will be left to the bidders to submit their interpretation and fiscal proposals. The purchaser must therefore recognize the need to clearly specify these commercial issues as neglect in this area may present some complex problems in determining who is liable and who eventually will accept the responsibility and the financial risks, which can be quite substantial. Some of these matters may be covered by the law of the country under which the contract is administered, or they may have to have the disputed issues settled by arbitration or by litigation.

The following financial provisions relating to a particular equipment procurement are discussed in this chapter:

1. *Contract price structure.* Contract price structure information for the supply and delivery of equipment and services of advisors for installation and commissioning.
2. *Contract price adjustment provisions.* Contract price adjustments to apply to variations in cost elements during the course of the contract.
3. *Terms of payment.* Terms under which all price payments are to be made to the contractor.
4. *Contract program.* Programmed milestone and completion period or dates to be achieved from date of award of the contract.
5. *Liquidated damages.* Contractor agreement to pay the purchaser as fixed, pre-estimated liquidated damages served for protection against late delivery and/or failure to meet the specified performance warranty.
6. *Insurance.* Coverage of defined risks.

6.1   CONTRACT PRICE STRUCTURE

The pricing information to be provided by the bidders for the work to be performed under the contract is generally set out in the form of fill-in schedules to be completed by the bidders.

The pricing information sought by the purchaser should clearly state those items needed for the evaluation of bids and for establishing the contract price. For an equipment supply procurement, the preferred contract price is a lump-sum. The lump-sum stipulates that the contractor must execute the work under the contract for an all-in price without any split or sub-division of component items forming the contract. As discussed in Chapter 1, a lump-sum price can be either fixed or subject to price variation, whichever is specified in the contract conditions. In this chapter, the contract contains cost adjustment provisions that are described in Section 6.2.

The lump-sum price normally covers the following elements:

- Cost of the equipment
- Transportation and handling charges for delivery to the nominated destination
- Taxes
- Import duties and tariffs
- Insurance charges as specified
- Services of the contractor's advisors for installation, commissioning, and personnel training, if specified
- Essential spare parts
- Unit price schedule for extra work that may be requested by the purchaser

The contract price may be either lump-sum fixed during the course of the contract and not subject to variation on any account, or subject to price variations arising during the course of the contract (see Section 6.2).

The contract price together with the defined trade terms[1] may be structured in one of two ways:

1. Contractor is responsible for supply only, purchaser is responsible for transportation:

    Price Ex Works, *or*,
    FCA, FAS, *or* FOB[1] $ ....

---

[1]See note on trade terms on page 6–3.

2. Contractor is responsible for supply and delivery:
   CFR - Cost and Freight
   CIF - Cost, Insurance, and Freight

	Equipment supplied within purchaser's country	Equipment supplied outside purchaser's country
Ex Works	$	$ ....
Transportation		
- Ocean		$ ....
- Land	$	$ ....
Insurance	$	$ ....
Customs duty		$ ....

Note on trade terms. In order to be consistent in this book with trade terms for both domestic and export transactions, the trade terms "Ex Works," "FOB," "FAS," and others used shall be defined by the current edition of Incoterms (see Appendix C) which provides a set of international rules for the interpretation of these terms. To ensure that bidders (and also the purchaser) understand these terms, it is advisable that the purchaser includes a copy of the definitions in the bidding and contract documents.

Bid Currencies

Prices indicated on the bid form are normally quoted in the following currencies:

1. The whole of the equipment supplied by the contractor from within the purchaser's country, in the currency of the purchaser's country.
2. Parts or the whole of the equipment supplied by the contractor from outside the purchaser's country, either in the currency of the contractor's home country or in U.S. dollars. The bidder shall indicate the currency for payment. Alternatively, the contractor may request part payment in home currency and part in U.S. dollars, indicating either nominated values or as percentages of the contract price.

For imported items, a major consideration is whether to accept the bidder's price in the foreign currency. Uncertain political situations and the effects of inflation on material and labor costs can greatly influence the selection for payments in either home or foreign currencies especially for long manufacture and delivery procurements.

Currency hedging is not dealt with in this text because it is a matter for the parties to take into account acting alone and unilaterally. It does not affect their relations one with the other and accordingly does not come within the scope of this book.

Taxes

It is important to specify matters relating to all taxes, stamp duties, and other such levies payable by the contractor on materials and equipment. Sales tax—variously known as consumption tax, goods and services tax, value-added tax—that varies from time to time within the country of manufacture, and in accordance with federal and state laws, and should be carefully considered.

It is a frequent practice that the contractor be entirely responsible for all such taxes imposed outside the purchaser's country but not be responsible for payment of such taxes imposed in the purchaser's country for imported materials and equipment. Where all the goods are supplied from the purchaser's country, the contractor is responsible for all taxes.

It should also be noted that in some instances, the purchaser may be exempt from paying a federal or state tax for equipment supplied covered by a particular contract that would otherwise apply under an applicable assessment law or act, and therefore should not be included in the contract price. Such tax exemption must be disclosed in the instructions to bidders document with reference made to the applicable exemption authorization to ensure that it is not included in the bidder's price.

Because of the complexity of tax-related matters, the purchaser should seek advice from an appropriate authority before specifying requirements in the bidding and contract documents.

Insurance

As insurance is a major consideration for both the purchaser and the contractor to the contract, it is the purchaser's responsibility to clearly set out the requirements for where he wishes the liability to be placed.

The purchaser will need to consider whether he is able to obtain a policy for the contract that is suitable for his needs, or to place the responsibility with the contractor (see Section 6.6).

The bidding documents should define each class of insurance clearly and concisely so that there is no confusion as to who is responsible, and to what extent coverage is required. Where limits apply, these should be stated. Where insurance is being covered by the purchaser, advice should be given that a copy of the policy can be viewed by the bidder. The bidder should also be advised of the timing of having an insurance in place when the contract is awarded, and any special requirements in relation to supplying copies of policies, and notice of renewals, or extensions during the life of the contract.

Import Duties and Tariffs

The bidder's contract price should include duties for imported equipment or separate parts thereof in accordance with tariff classifications. As the duty rates can be changed without notice from the time the bid is submitted to the time payment is claimed, it is essential that bidders include the following information for each item imported on a separate schedule supplied by the purchaser:

- Name of item imported
- Country of origin
- FOB value that prevailed at the bid nominated base date of the contract
- Currency exchange rate at bid nominated base date of the contract
- Customs tariff item
- Customs duty rate

Care should be taken to check that the named piece of equipment is not confusing and thereby cause a wrong tariff item to be used and charged for by the customs department. If this remains unchecked, the purchaser can be subject to unnecessary duty payment.

Services of Contractor's Advisors (if specified)

Included in the lump-sum contract price but to be quoted separately, are the services of the contractor's personnel for on-site activities. This will permit the purchaser to negotiate with the favored bidders, both the estimated number of days for personnel to be in attendance at the site and also the unit cost for any additional services that may be requested by the purchaser. The price to be quoted is for .... (*number*) of calendar days at the site from the time of departure from the home station to the site and return.

The purchaser will normally pay for:

- A unit rate including overheads
- Travel expenses from home station to site and return
- Necessary daily transportation expense
- Meals and accommodation

Spare Parts

It is desirable to have spare parts available when needed as soon as the equipment commences commercial operation. It is more economical to procure these parts at the same time as the equipment is being manufactured. The scope of spare part requirements is not known when soliciting bids and it is normal for bidders to include in their bids a list of parts as follows:

- *Essential.* Spare parts to be supplied as part of the contract as specified in the technical specification, the value of which is included on the contract price.

- *Optional.* Spare parts recommended by the contractor for a specified period of time, each item quoted on a unit price basis, inclusive of sales tax where applicable, for supply and delivery to the purchaser's store.

Unit Price Schedule for Extra Work Requested by the Purchaser

In order to avoid negotiating prices for extra work requested by the purchaser after award of contract, the bidder should be required to prepare a separate schedule detailing rates for the following:

- Unit rates charged for home office personnel
- Unit rates charged for site advisors
- Travelling and subsistance
- Categories of billable and nonbillable personnel
- Printing and reproduction service charges
- Computer and data processing service charges
- Telephone and facsimile charges

The above rates should also apply to the contractor's subcontractors.

The purchaser should ensure that evidence of these services will be forthcoming on request before making payments.

6.2   CONTRACT PRICE ADJUSTMENT PROVISIONS

Of major importance when formulating the contractor price is for the purchaser to determine whether the contract is to be on a fixed price basis, implying that the contractor is not entitled to claim any increase in the contract price for variations in the cost of labor, materials, transport, and other major costs during the life of the contract. For a short-term procurement, the contractor is usually able to control or absorb variations in the cost elements. However, for long-term procurements, bidders must take into account economic factors and anticipated courses of inflation, resulting in an instability of costs. Bidders will then be requested to quote a base price that will be adjusted upward or downward to compensate for increases or decreases in cost elements over those in effect at the bid closing date. Such price adjustments will protect the contractor's financial position should there be major increases in either or both labor and material costs during the life of the contract.

Foreign currency exchange and customs duties are required to be subject to price adjustment for fluctuations in costs that occur after a specified base date stated in the contract. The bidding documents should state that bidders are to indicate in the bids which cost elements of the contract price are to be subject to price adjustment for such fluctuations in costs. They normally include:

- Labor and material costs
- Transportation and handling costs

The amounts of price adjustment should be based on changes in the cost of the major components of the contract. The supplementary conditions should provide escalation formulae for these cost variations so they can be calculated at the time when payments are due to the contractor. The cost variations are normally based on the official movement of price indexes published at regular intervals by national sources.

Labor and Material Cost Variations

If, by reason of any rise or fall in the rates of wages payable to the contractor's personnel engaged on work under the contract or in the cost of material above or below such costs ruling at the bid closing date (or the base date[2] stated in the contract), the cost to the contractor is increased or reduced and the amount of such increase or reduction is added or deducted from the contract price as the case may be.

Variations in costs are made in accordance with escalation formulae based on indices or factors published at regular intervals by a specified federal government department or statutory agency responsible for compiling such statistics. Methods for determining the applicable indices or factors must be clearly defined. This information is preferably detailed in the purchaser's supplementary conditions and is subject to acceptance or negotiation by the selected bidder prior to the award of the contract. Alternatively, bidders are permitted to submit their proposed escalation formulae but this procedure will greatly extend the bid evaluation process.

For the purpose of adjusting the contract price, a specified percentage of the contract base price is to be considered as the labor cost.

The material cost can be either a specified percentage of the contract base price for calculation in the escalation formulae, or be adjusted by an amount equal to the increase or decrease in the contract base price that the contractor is obliged to pay. It can be calculated in a prearranged manner, for example, for every one cent per kilogram of raw material change in the base contract price, the contract base price shall be altered by $...................... (*state amount*).

---

[2]The base date is the date used to define the index for labor and materials that was used to prepare the bid for these items and values subject to contract price adjustment provisions. It is recommended that a provision in the contract conditions should read:

The base date of the contract for price adjustment provisions shall be 28 days prior to the bid closing date.

In selecting the percentage to apply to labor and material, care should be taken to ensure that the bidder does not show unrealistic percentages so as to obtain a greater increase than he is entitled to due to one factor receiving higher increases than the other.

Ocean Freight Variation

Ocean freight rates ruling at the bid closing date shall be stated by the bidder. Any variation in the freight cost is normally calculated in the direct proportion of the actual freight rates at the date of dispatch of the freight to the freight rates at the stated bid closing date (or the base date). Any increase or reduction in such transportation costs shall then be added or deducted from the contract price.

6.3   TERMS OF PAYMENT

The terms of payment should be set out in the supplementary conditions. While these are subject to negotiation with the favored bidders during the bid evaluation process, it is wise for the purchaser to specify the preferred terms in the bidding documents. Naturally, the purchaser will try to avoid committing large payments before the project is capable of earning revenue. Likewise, no contractor wishes to strain his cash position having to finance the work up to delivery to cover material purchases, labor, and other charges without receiving payment to offset these expenditures.

To protect both the purchaser's and contractor's financial interests, the contract should make reference to the following for the supply and delivery of the equipment (installation by others):

- Define precisely the milestone event in the contract when payment becomes due to the contractor, such as:
  - Ex Works
  - FCA, FOB, or FAS
  - Delivered at the point of destination (the project site)
  - Taking over the equipment
  - End of the equipment warranty or defects liability
- State the amount due at each event, or provide a means by which such amounts of payment can be determined
  - A percentage of the contract price together with any price adjustment
  - Progress payments made during manufacture
- Establish a time limit within which payment must be made to the contractor, and the course of action should the purchaser default in payment

- Holding back a certain percentage of the contract price (retention money) until the end of the equipment warranty or defects liability

Examples of how payments can be made at various stages during the life of the contract are shown in Figure 6–1. The examples are not exhaustive as many other variations can apply. The figures shown are a percentage of the contract price.

Option	1.	2.	3.	4.	5.	6.
With order		10				
Ex Works	90*					60*
FCA[1], FOB[1], FAS[1]			90	85	90	30
Delivered at site[1]		80		5		
Delivered at site[1] and Taking-over	5		5			
Taking-over[2]		5		5		
Retention[3]	5	5	5	5	10	10

*Progress payments made in equal increments or on progress of the work
[1]On presentation of valid shipping documents
[2]After the equipment successfully met the on-site tests
[3]Retention based on the contract terms

Figure 6–1   Options for payment during the life of the contract.

Method of Application for Payment

Application for payment can be made in the following methods:

Payments up to and including those payable on dispatch from the manufacturer's facilities. Payment should be made following the receipt of payment certificates prepared by the contractor and approved by the purchaser. Payments due to the contractor should be made within .... days specified in the contract conditions after receipt of the certificates.

Payments for transportation and insurance (if applicable). In respect to equipment shipped to the project site, full payments should be made as set out in the contract price schedule within .... days specified in the contract conditions, after receipt by the purchaser of proper evidence, properly certified showing that transportation and insurance charges have been paid in full. If not all of the equipment is shipped together, then pro-rated payments may be made on partial shipments.

Customs duties. Should be reimbursed by the purchaser based on actual payments made by the contractor on conditions described above.

Payments for services of the contractor's advisors. Should be made in accordance with the provisions set out in the contract documents.

Payment for extra work authorized by the purchaser. Should be made on the contractor furnishing evidence of costs incurred and in accordance with the provision set out in the contract documents.

Letter of Credit

As many procurements involve importing equipment from overseas, the selected bidder may request that the contractor (if the bidder is not the prime contractor) be paid for the exported components by *Letter of Credit*. This method of payment is subject to negotiation before award of contract. If the arrangement is satisfactory to the purchaser, then the terms of the letter of credit arrived at by negotiation would become part of the contract document.

A letter of credit provides a method by which the contractor who is exporting his product is able to claim payment from a bank that he nominates in his own country, on presenting that bank with evidence of having completed his contractual obligation, such as bills of lading, or whatever other documents are required by the terms of the letter of credit. It is a guarantee that the contractor will be paid by the bank if he complies strictly with the terms of the letter of credit.

The procedure for establishing a letter of credit is as follows. When a contract agreement has been signed, the purchaser instructs his bank to set up a letter of credit by informing the bank of the precise terms of payment contained in the contract. The bank, when undertaking this obligation on behalf of the purchaser, may demand a security such as a deposit of part or all of the cash required. Approval to establish the letter of credit may also be required under the foreign law. The purchaser's bank will then contact the bank in the contractor's country that, on accepting the advice note, will issue a confirmation to the contractor of the establishment of a letter of credit.

It is important to note that the contractor's bank is required to carry out the terms of the letter vigorously. This applies particularly where the contract provides for contract price adjustments (escalation) during the course of the contract, and the letter of credit must make reference to this if the contractor is to be paid without delays.

Retention Money

If at the time when the last payment is due to the contractor at the end of the equipment warranty or defects liability period, there are defects in the equipment that are considered to be minor and will not affect the commercial operation of the equipment, then the purchaser is entitled to retain only such part of the last payment as represents the cost for rectifying such minor defects. Any sum of money retained by the purchaser will be paid to the contractor once the defects have been rectified. In lieu of the retention money held, it is frequent practice for the contractor to provide the purchaser with a surety bond or bank guarantee for the amount of retention money (see Chapter 13).

6.4   CONTRACT PROGRAM

Each contract has certain milestone dates that are not only of legal importance but become the spine of the contract since they bind both the purchaser and the contractor for the successful execution of the work. These milestone dates form the *Contract Program.*

The contract program contains those defined dates that the contractor must meet and also certain defined dates the purchaser is obliged to meet, such as making available specified information or facilities the contractor needs during the course of the contract. In essence, the contract program is the measure of the contract.

The contract program has legal implications for both the purchaser and the contractor because the contract program establishes the legal enforceable dates. These dates are usually used for application of liquidated damages for delays in meeting milestone intermediate dates and the contract completion date. Also any claims of extension of time will need the contract program dates to be in place so that new dates can be applied.

The contract program should not be confused with the work program, which the contractor may be obliged to produce for the purchaser to show the program for the principal activities of work he will perform (design, ordering materials, etc.), and the programmed starting and completion time (or date) for each of these activities. The program is mainly used by the purchaser to monitor the contractor's progress. Unless noted otherwise in the contract, the contractor is not contractually obliged to adhere to the activities noted in the work program, provided the contractual dates are met, which may include other milestone dates that the purchaser requires the contractor to achieve and that are specified in the contract program. The following are key contractual dates that must be clearly defined in the contract documents.

The contract date. This is the date when the contract comes legally into existence. It is usually the date of the purchaser's notification to the successful bidder of the acceptance of the bid.

The effective starting date of the contract. This is the date when work on the contract actually commences. It may coincide with the contract date but could also be a later date specified by the purchaser. Examples for a later date for starting the work include a design feature to be supplied by the purchaser is incomplete, or where there is a delay pending receipt of a statutory permit. If the effective starting date is not the contract date, it should then be specified in the contract documents and used as the base date for the calculation of the contractual period for equipment delivery or contract completion.

The contract completion date. This date is of considerable significance and has already been discussed in Chapter 5, page 5–14. As noted, the completion date should be clearly defined in the contract conditions and the actual date stated in the supplementary conditions. Should the contractor fail to complete the work by the specified completion date or by any extension of time granted by the purchaser, he may be subject to liquidated damages for delay in equipment delivery or on-site completion.

6.5   LIQUIDATED DAMAGES AND BONUS PAYMENT

Provisions relating to the contractor's liability and the purchaser's claims for liquidated damages are usually set out in the general (standard) contract conditions as discussed in Chapter 5. The purchaser's specific requirements relating to a particular contract are incorporated in the supplementary conditions. They apply to the contractor's failure to meet the contractual delivery date and/or the specified equipment performance warranty. To compensate the purchaser, the magnitude of the damages in the form of specified monetary payments should be based on the financial value of the lost production and only be invoked by the purchaser in the event of actual losses having been incurred.

Where the purchaser wishes to pursue this course of action, the liquidated damages provisions must specify the amount envisaged for compensation. Incremental periods of delay of delivery and/or failing to achieve the specified minimum warranted equipment performance together with the sums to be claimed must be accepted by the contractor during the bidding stage.

Care must be taken by the purchaser to ensure that the liquidated damages do not exceed a fair estimated value of what he would be expected to suffer from the delayed delivery or the equipment failing to achieve the specified performance.

Any greater values would be regarded as a form of penalty imposed by the purchaser against the contractor and such claim may not be legally permissible. However, courts generally support the purchaser's use and enforcement of a liquidated damages provision if the amount is reasonable and not punitive.

The contract must also state quite specifically the minimum performance to be achieved whereupon the purchaser will accept the equipment.

To ensure equitable administration and assessment of liquidated damages, certain considerations should be taken into account by the purchaser. They include:

- Establishing acceptable performance figures
- Establishing a reasonable contract time for equipment delivery
- Stating the starting date of the contract in the contract documents
- Clearly defining the meaning of the delivery date in the contract documents
- Establishing a monetary amount of liquidated damages that has been arrived at reasonably

The total liability of the contractor for the agreed pre-estimated and liquidated damages for delayed delivery of the equipment normally does not exceed 5 percent of the contract sum. The amount claimed by the purchaser is deducted from the contract price.

6.5.1   Delay in Delivery

Liquidated damages for late delivery to the site are usually calculated on an amount of money per day of the contract price for every day that has elapsed after the base date of contract until all of the equipment has been received at the site.

As a guide for a supply and delivery procurement, the liquidated damages provision in the contract conditions may read:

> In the case where the contractor fails to deliver the .... (*state name of equipment*) on or before .... (*state the date specified in the delivery schedule or any extension thereof*), the contractor agrees to pay the purchaser as fixed, agreed pre-estimated and liquidated damages and not as or in the nature of a penalty, the sum of $...................... (*state sum of the money*) for each calendar day of delay beyond the number of days stated in the delivery schedule.

It is not unusual to limit liquidated damages to a percentage (for instance, 5 percent) of the contract price, otherwise bidders will load their prices to cover any unforeseen event.

6.5.2   Equipment Performance

Liquidated damages can also be drawn upon should the equipment fail to meet a minimum specified performance requirement. These damages may apply to contracts where the equipment:

- Fails to meet guaranteed efficiency
- Fails to produce the guaranteed production rate or capacity
- Fails to handle the guaranteed amount of material

The above examples will result in increased operating costs to be borne by the purchaser over and above the costs had the specified warranted performance been achieved. The liquidated damages are therefore computed on the basis of the losses the purchaser will sustain and capitalize over the life of the plant and equipment.

A sample provision for liquidated damages may read:

> Should the .... (*state name of the equipment*) output or efficiency (*state what is applicable*) when performance tested, be less than .... (*what is specified*) at the conditions set out in .... (*state applicable document and clause*), and should the contractor fail to alter the equipment within a reasonable time so it will meet the warranted performance, and should the equipment be otherwise than in a satisfactory operating condition, the purchaser in lieu of rejecting the equipment may accept it conditionally and in that event, the contractor agrees to pay to the purchaser, as fixed, agreed preestimated and liquidated damages and not as or in the nature of a penalty, the sum of $..................... (*state sum of the money*) for each .... (*state condition, e.g., percentage, or tonne production, or* ....) below the contractor's warranted performance.

6.5.3   Bonus Payment

Likewise with the imposition of liquidated damages, the contract documents should clearly indicate the payment of bonus, if any, to the contractor for the achievement of early completion of the work by the target dates set in the contract program. However, a purchaser may prefer to negotiate a bonus for a better than normal delivery with selected bidders after bids have been evaluated and before award of contract, since he would only be interested in offering an increase in the contract price for what it is worth in achieving results that would benefit the production program. It should be noted that under U.S. law, penalty provisions can be illegal without corresponding bonus provisions for better than specified performance.

6.6   INSURANCE

Insurance of the contract work is a matter of considerable importance to the purchaser. It is the purchaser's responsibility to ensure that both he and the contractor are sufficiently protected against any exposure that could place the purchaser at risk, and that the insurance will make good such loss and any subsequent loss resulting from a primary loss.

The business of insurance is highly complex and specialized and should be approached with extreme caution. It is imperative therefore that the purchaser's engineer seeks the assistance of his or her company's insurance advisors on all matters relating to insurance before drafting the bidding documents. It is the intent of this book to go no further than to acquaint the engineer with the basic principles of insurance and to create an awareness of some of the problems that can arise due to the unfamiliarity of this important section of a procurement.

In general, an insurance policy is an agreement whereby one party (the insurer or insurance company) undertakes for a stipulated consideration (the premium), to indemnify (compensate) the other party (the insured) against loss arising from the particular peril that is specified in the insurance policy. It is a means of transferring part of the "risk" or liability from the insured to the insurer. The insurance policy itself is a legal document that sets forth the agreed limits of liability both in terms and consideration for which the insurer has assumed responsibility.

Agreement between the insured and the insurer is always subject to government insurance codes and other statutory provisions. The insurance provisions to be drafted in the bidding documents should reflect the type of procurement to be provided under the contract, that is, FOB, CIF, or otherwise. Provisions defining the contract insurance responsibilities so as to ensure adequate cover is provided against potential legal liabilities are usually set out in the contract conditions.

Whereas adequate cover is always necessary, doubling up of insurance inferring that the same risk is covered by two different insurance policies should be avoided. It is therefore frequently in the interest of both the purchaser and the contractor that a single insurance company deals with any *class* of claim.

Apart from insuring the delivery of equipment to the site, which is discussed separately, it is assumed in this book that the site is already insured by the purchaser; this will include installation work.

It becomes most economical and practical then for the purchaser to incorporate the startup work for the duration of the defects liability period into this existing policy. If the purchaser elects to provide this insurance cover, a copy of the policy should be included as an attachment to the contract conditions. This will enable the contractor to check whether additional cover is required. Alternately, should the purchaser require the contractor to provide the insurance cover, this must be clearly defined in the contract conditions. This could occur for public liability and if manufacturing takes place at site. In any event, it is recommended that the purchaser be named a co-insured on any contractor or subcontractor policy.

Regardless of whoever is responsible for the insurance cover, the names of the insured stated in the insurance policy should be the purchaser, the contractor, and subcontractor(s) for their respective rights and interests and liabilities. The policy, also known as the "contract insurance," should state:

- Name of the contract
- Risk location

- The insured property
- The period of insurance
- Limits of liability with provision for escalation
- Amount of deductible or excesses
- Adjustment of premium
- Currency for payment of premiums and losses
- Basis of settlement
- Any exemption clauses
- Exclusions

The following are the main areas to be covered by insurance:

- Workers' compensation—insurance of the contractor's and the purchaser's employees against death, bodily injury, or disease
- Transit insurance of the equipment to the site
- Equipment while on the site
- Public liability against claims by third parties

Workers' Compensation Insurance

It is assumed that the purchaser is fully conversant with the insurance of employees. Compensation for death, accidental personal injuries, and certain occupational diseases to an affected employee (whether permanently employed or casual) arising out of and in the course of employment is mandatory by law in most countries. Nevertheless, the purchaser should ensure that the contractor has insurance cover against this liability whether by law or not, for death, injury, or occupational disease for his employees as well as all subcontractors engaged on contract work. Cover under a workers' compensation policy is frequently unlimited with the law of some countries setting a minimum sum cover for any one occurrence. This is to be maintained until all work is complete, including the end of the defects liability period. Where this is not taken out by the contractor or is inadequate, the shortfall in liability could fall to the purchaser.

Transit Insurance

The responsibility for the transportation of the contract equipment to site and the basis of delivery is set out in the bidding documents, for example, Ex Works, FOB, or complete delivery to the site. Should all the work be insured by the purchaser, the insurance during shipment should not be included by the contractor or provided for in the schedule of prices. Consequently, the contractor will not be responsible for loss, damage, or depreciation of the equipment unless it is due to faulty protection or insecure packing.

The modes of transportation for the purpose of insurance are:

- Overland including loading, unloading, transshipment, and storage en route
- Transportation by sea or air including all risks of loss destruction or damage during loading, unloading, transshipment, and storage en route

Terms of transportation insurance cover are subject to terms, conditions, and exceptions contained in numerous insurance institute clauses current at the date of shipment. They include, just to name a few:

- Institute of Cargo clauses
- Institute of Air Cargo clauses
- Institute of War clauses
- Institute Replacement clauses
- Deck Cargo clauses

Insurance at Site

Apart from workers' compensation, the principal objects of insurance are to ensure that loss or damage to any section of the contract work is sufficiently covered by insurance before take over or passing the operating risk to the purchaser no matter how caused.

Exceptions to the risk by the insurer usually include:

- Normal wear and tear
- Loss or damage arising from war, invasion, nuclear radiations, and the like
- Loss from the consequences of civil war, riots, and civil commotion

It is customary practice to show in the policy the contract price inclusive of the cost value of the purchaser's free issue of material and services, customs duties paid, transportation costs, and insurance. The amount insured or replacement costs should not be less than the contract price with additional indemnity for:

- Demolition and removal of wreckage
- Any temporary work as required
- Loss of, or damage to construction plant at the site
- Engineering expenses incurred from payment of professional fees in connection with restoration work
- Additional expenses incurred for expediting and delivery of plant and equipment for the reinstatement of the loss
- Miscellaneous expenses incurred on behalf of the insured

The policy should be for full cost and include automatic adjustments for cost escalation so as to reflect the full replacement value of the work at all times.

Deductible limits or excesses are amounts stated by the purchaser in the contract conditions that are to be borne either by the contractor or the purchaser, and depend on local practice.[3] Care should be taken when determining the amount of excess. Setting a high excess will mean a reduction in premium but will also increase the extent of a claim.

Where the contractor takes out the insurance, the purchaser should carefully note any deductibles or excesses as they can be a guide to the contractor's past insurance claim history.

Public Liability and Property Damage Insurance

This insurance protects the purchaser's legal liability to pay compensation to a third party for:

- Bodily injury suffered or alleged to have been suffered by any person or persons
- Loss of, and/or loss of use of, destruction of, or damage to any property. Depending on the cover of the policy, property may also include property owned by the insured other than where contract work is taking place.

Limits and exclusions normally include:

- Employees of the contractor, subcontractor(s), and of the purchaser who are normally covered by the workers' compensation insurance
- Liability arising out of the rendering or failure to render professional advice
- Loss or damage from causes other than planned demolition work or similar risks
- Consequences as a result of an explosion or rupture of a fired or unfired pressure vessel or storage tank not part of the contract
- Rectifying faulty design, material, or workmanship on contract work
- Acts of God (force majeure)

If the contractor is required to cover the public liability and property damage risk, it should not be for less than the amount the purchaser has stated in the contract conditions. This amount will not affect the contractor's liability and it is advisable for the purchaser to fix a realistic sum for the amount of insurance, taking account of possible claims and realizing the amount of insurance paid for will affect the premium to be paid and passed on in the contractor's bid price.

---

[3]Certain international standard practices require the contractor to pay any deductible or excess if the purchaser carries the insurance. It appears that current U.S. practice is for the purchaser to pay any deductible if he carries the insurance.

The public liability and property damage insurance should be maintained until the completion of the defects liability period.

Finally, the contract conditions should include provisions to cover:

- That the purchaser must be satisfied at all times that the contractor has maintained adequate approved insurance cover. The contractor must produce insurance policies together with receipts on demand by the purchaser. The purchaser may be unaware that the contractor has cancelled or failed to renew a policy, leaving the purchaser without insurance cover. A record of renewal dates should be kept and confirmation of renewal requested.
- A proposed change by the contractor to any insurance policy must first have the purchaser's approval.
- If the contractor fails to maintain a stated policy, the purchaser may do so on the contractor's behalf and deduct the cost of the premium from the contract price. This is usually included as a provision in the contract conditions.

Exclusions

The exclusions mentioned above as frequent standards for the policies discussed may usually be included as insured risks by negotiation with the insurer and payment of an additional premium.

7

PREPARING THE TECHNICAL SPECIFICATION

7.0 INTRODUCTION

The most detailed part of an engineering contract is the technical specification. It is the tool that details the technical aspects of the procurement contract for equipment and services to be furnished in accordance with the purchaser's requirements. The technical information and requirements of the work must be clearly understood and interpreted by not only professional engineers but also persons coming from widely differing backgrounds. It must therefore be written quite explicitly and in plain, simple language.

A thorough knowledge of the equipment requirements and approach to the various work activities is essential. Coordination with groups within the purchaser's organization is usually required for seeking sources of information needed for preparing the documentation. The effort demands extreme care to comprehend and accurately describe the requirements and avoid inconsistencies, omissions, and duplications. Afterthoughts and corrections having to be made later in the contract can involve expensive and wasteful redesign, reworking, and delay.

For major equipment procurements, the technical specification comprises a number of separate engineering documents. These documents will vary in scope and detail depending on the type and size of equipment being procured but usually include most, if not all, of the following:

- Technical requirements form
- Engineering specifications
- Equipment data sheets
- Reference drawings
- Miscellaneous technical documents related to the procurement
- Contractor's drawing and technical data submittal requirements

For developing the documents, it has been assumed that an equipment design criteria has been completed and approved for the procurement to proceed. Major items in the criteria would concern operating and performance characteristics for achieving scheduled production rates and quantities.

The summary of the technical specification is accomplished within the framework of a single document titled Technical Requirements Form, discussed in Section 7.1.

7.1   TECHNICAL REQUIREMENTS FORM

The *Technical Requirements Form* is an engineering form completed by the purchaser to tabulate what equipment and service, if any, the contractor is to supply. The form together with the documents listed thereon, comprise the technical specification for the procurement. An example of a technical requirements form is shown in Figure 7–1 (page 7–4).

This one document should be used to cover the requirements of the technical specification in preference to having a series of separate documents, either bound or loose, identified by title only, since the technical requirements form will identify whether any one or all documents have been revised since their initial issue. Furthermore, it is a checklist of all the documents, each with its current revision status, that form the technical specification to be used by the purchaser's personnel engaged with the engineering and purchase (including expediting and inspection), as well as by bidders and the contractor.

The form is a working document from the bidding stage through to the completion of the procurement. It should be noted that the form and the listed documents are technical documents and therefore should not make reference to price, delivery, or other commercial matters.

7.1.1   General Format

The form is divided into the following four sections to be filled in by the purchaser:

1. General information block
2. Attachments block—listing technical documents
3. Equipment information block
4. Revision block

The principal contents of each block are described below.

General Information Block

- Title of the equipment being procured
- Contract number
- Project number
- Revision number (or current status of the document)

Attachments Block—List of Technical Documents

This block, titled "Attachments," lists all the technical documents that are included with the technical requirements form. As shown in Figure 7–1, Item 1, "Drawing and Technical Data Submittal Requirements Form," is preprinted on the form since it is a document that would be issued for every equipment procurement.

The other attachments normally include:

- Engineering specifications
- Equipment data sheets
- Purchaser's drawings—designated as "reference drawings"
- Miscellaneous technical documents relevant to the procurement such as reports, or chemical and physical analyses.

Equipment Information Block—Item Number, Quantity, and Description

The item number, quantity, and a description of the equipment being procured. Example:

Item 1	1	Fire water pump with electric motor driver Tag Equipment: 1201
Item 2	1	Fire water pump with diesel engine driver Tag Equipment: 1202

Revision Block

This block contains the following:

- Revision status
- Date of issue
- Description of issue, such as:
  - Issued for bids
  - Issued for purchase
  - Revised as noted
- Initials of persons issuing and approving the document

7.1.2   Issuing and Revising the Technical Requirements Form

When the technical requirements form is first issued for soliciting bids, it is assigned Revision 0 (or simply Rev. 0), and in the revision block is designated "Issued for bids." A technical requirements form completed and issued for bids for a sample procurement is

XYZ CORPORATION

TECHNICAL REQUIREMENTS FORM

TITLE _____*(name of equipment)*_____   CONTRACT No _____

_____

_____   PROJECT No _____

DELIVER TO _____*(name of project)*_____   REVISION No _____

_____*(address)*_____

_____   SHEET No __1__ OF ____

_____

ATTACHMENTS

	THE FOLLOWING DOCUMENTS FORM PART OF THE TECHNICAL SPECIFICATION	REV
1	Drawing and Technical Data Submittal Requirements Form	
2		
3		
4		
5	*(list all technical documents with current revision number)*	
6		
7		
8		
9		
10		

REV No	ITEM No	QTY & UNIT	DESCRIPTION	ACCOUNT No
			(equipment to be procured with applicable	
			equipment tag number)	

REV	DATE	ISSUE DESCRIPTION	BY	APPROVALS

FIGURE 7–1   Example of a technical requirements form.

shown in Part 5, Section B.1. Initial in-house issues for review or document approval prior to release for bidding are in the series "A," "B," "C,". . .

Should a modification be made to any document listed in the attachments or description block during the bidding period, the form has to be revised and is reissued to all bidders as Rev. 1, designated "Revised as noted and reissued for bids." A revised form for the sample procurement is shown in Part 5, Section C.3.

Upon completion of the purchaser's bid analysis and the selection of the successful bidder, the form is revised to include all technical documents that have been brought to an "as purchased" status as discussed in Chapter 12. Documents listed in the attachments and issued solely for bidding purposes are now deleted from the form (not erased but deleted by means of a simple ruling-through line). Where modifications and updates have been made to engineering specifications, data sheets, drawings, and other technical documents, each of these documents will have the respective updated revision number noted in the attachments. When finally issued with the contract documents, the technical requirements form is assigned the next higher revision number, Revision 2 (or Rev. 2), for the procurement and designated "Issued for purchase."

The technical requirements form and the attachment documents are distributed to the purchaser's personnel, to bidders when soliciting bids, and to the contractor when the equipment has been committed for procurement. The form and the respective documents will be reissued each time a revision is made, and the form itself remains in use until completion of the contract, revised whenever reissued.

7.2   TYPES OF ENGINEERING SPECIFICATIONS

Specifications describe in detail the equipment and services required from the contractor. They are usually accompanied by data sheets(s), which specify details of operating, performance, and construction parameters. Two categories of engineering specifications are described in the following paragraphs.

7.2.1   Equipment Design and Performance Specification

An *Equipment Design and Performance Specification* is normally a one-of-a-kind specification, written for a particular project and directed to contractors (or equipment suppliers). It will define the type of equipment, materials, or services that are desired for the components or for a complete system. The specification may also be supplemented by applicable standard specifications as described in the next section. Other applicable reference documents often include standards, codes, and regulations, which are described in Section 7.3.

7.2.2   Standard Engineering Specification

Every specification writer seeks ways to ease the task of writing the document. One obvious technique is to copy a previously written specification making appropriate changes. This option requires extreme care, since otherwise inconsistencies will be incorporated. A preferred method would be to use an applicable *Standard Engineering Specification* that is already available to the writer. A standard engineering specification is a technical document that has been prepared and written as a technical specification without reference to a particular project.

Many engineering organizations, both private companies and government authorities, have prepared a series of standard specifications for the type of work likely to recur in their businesses. In most instances, these standard specifications have been developed over many years of usage and cover a wide range of equipment, facilities, and services. The most common would include:

- General data and requirements applicable to most projects
- Piping, valves, and fittings
- Electric induction motors, power transformers, and the like
- Platforms, walkways, ladders, and stairways
- Painting, insulation, and other material applications
- General noise control

With the ready availability of word processing software, master copies of these specifications are conveniently kept on disks and can be used when and as required. Having these documents on record for repeated use eliminates the time-consuming task of rewriting, checking, and correcting the same specifications each time a new project is undertaken.

A standard specification is only revised or updated as and when required to reflect changes in engineering practices, procedures, and/or techniques, and to update standards, codes, and statutory regulations.

It should be noted that if a standard engineering specification needs to be modified, rather than revised, to suit specific contract requirements, the document is no longer identified as a standard specification and it becomes a project specification for the particular contract.

7.3   REFERENCE STANDARDS, CODES, LAWS, AND REGULATIONS

The work to be performed under the contract must be carried out in compliance with current editions of applicable standards, codes, laws and regulations.

7.3.1 Standards

A standard as defined by the International Organization for Standardization (ISO) is a document, established by consensus and approved by a recognized body (usually a national standards association), that provides for common and repeated use, rules, guidelines, or characteristics for activities or their results, to achieve the optimum degree of order in a given context.

A standard will prescribe a set of rules, conditions, or requirements concerned with the:

- Definition of terms
- Classification of components
- Delineation of procedures
- Specification of dimensions, materials, design, performance, or operations
- Measurement of quality and quantity in describing materials, products, systems, or practices
- Descriptions of fit and measurement of size

In most countries, standards are usually prepared by sponsored committees. They are published as National Standards by a national standards authority, for example, BS in Britain, DIN in Germany, JIS in Japan, AS in Australia. In the United States, these may be prepared and issued by other than a government entity. However, the American National Standards Institute (ANSI) is the only approving authority in the United States for national standards. It should be clearly understood that ANSI does not write standards. These documents are prepared by engineering institutions, industry-based organizations, trade associations, and individual companies. Only those standards that carry a designation ANSI/..................... are approved as American National Standards, for example, ANSI/IEEE 113-1985, Test Procedures for Direct-Current Machines, where the writing organization was the Institute of Electrical and Electronic Engineers.

7.3.2 Codes

Codes are usually enforceable at law and are written in mandatory language, i.e., "shall" in lieu of "should." They cover such equipment as boilers and unfired pressure vessels, cranes, hoists, and lifts (elevators). Major engineering societies and institutions in the United States, such as the American Society of Mechanical Engineers (ASME), have published numerous safety and performance test codes for most major equipment and are ANSI approved. Similar codes have been published in other countries.

7.3.3   Laws and Regulations

Individual laws pertaining to equipment procurements are usually not quoted in engineering specifications, although a paragraph must be included stating that the work furnished under the contract must conform to all applicable laws having jurisdiction at the purchaser's site.

Regulations are binding legislative rules published by government authorities and are mandated at law. They are issued at federal, state, and local government levels and where applicable, the work under the contract must conform to the requirements of these documents.

In the United States, federal regulations are contained in the Code of Federal Regulations and cover a very extensive system of regulations. Regulations at federal and state levels for equipment procurements vary from one country to another, especially those relating to environmental protection, mine safety, and occupational health. Local governments are mainly concerned with the establishment of uniform building regulations.

Laws as well as regulations at state level often make reference to codes and standards. For example, fired and unfired pressure vessels in almost all states in the United States are required by law to conform to the ASME Boiler and Pressure Vessel Code. Similar legally mandatory requirements are practiced in other countries.

7.4   WRITING ENGINEERING SPECIFICATIONS

The engineering specification must always be clearly understood and therefore, it must be written quite explicitly and in plain simple language. Completeness, clarity, accuracy, neatness, detail, and brevity are therefore essential considerations. The composition should be logical, systematically presented and be complete in scope and detail. Matters of major importance must be adequately covered to avoid misunderstandings while those of lesser importance may warrant only a brief explanation. However, nothing of relevance can be omitted.

Careful consideration should be given to the quality of the equipment desired. First and foremost, normal and rated performance requirements must be satisfied. However, it should be realized that depending on the type of service, the most efficient or most expensive equipment may not necessarily be needed.

For the type of procurements dealt with in this book, consultation with prospective manufacturers is highly desirable to avoid producing a specification requiring too many exceptions, or generating no response to the request for bids, thereby defeating the main purpose of the specification.

7.4.1   General Format

Because of the wide variety of engineering work encountered, numerous systems and techniques have been used to draft engineering specifications. In spite of this, it is possible to develop certain basic principles that cover the general format of specifications for most equipment procurements.

The decision as to the appropriate format will depend on the type, size, and nature of the equipment to be procured. The specification should preferably be organized so that the primary information is presented first and the secondary information is presented later as shown below.

- General statement of the scope of the work covered in the specification
- Work to be performed by the contractor
- Work to be performed by others
- Equipment design and performance requirements
- Technical requirements for the construction of the equipment
- General requirements

Specification Title and Continuation Sheets

The engineering specification should preferably be presented in two standard forms.

The *Specification Title Sheet* shown in Figure 7–2 (page 7–10) is sheet 1 of the specification and notes the following information:

- Name and address of purchaser's company
- Project name and number
- Title of the specification
- Specification number
- Contract number
- Sheet 1 of the total number of specification sheets
- Revision block showing:
  - Revision or issue number
  - Issue date
  - Issue description
  - Initials or preferably the name of originator and approving person(s)
- Table of contents for a multipage specification is optional

XYZ CORPORATION
2547 TENTH STREET
ANYTOWN NJ 00001

(Name of Project)

Project No _____

ENGINEERING SPECIFICATION
No _____

(Title of Equipment)

CONTRACT No _____

Sheet 1 of ____

ISSUE	DATE	ISSUE DESCRIPTION	BY	APPROVALS

FIGURE 7–2 Example of a specification title sheet.

An example of a *Specification Continuation Sheet* is shown in Figure 7–3 (page 7–12). Each continuation sheet has a title block noting:

- Name of purchaser's company
- Title of the specification
- Specification number
- Contract number
- Respective sheet number of the total number of specification sheets
- Revision number
- Initials of originator and approving person(s)

Specification Sections

The specification format should be organized by section and subsection. A *section* usually denotes a basic unit of work and receives a sequential number. A *subsection* is a subdivision of the section and each is designated by decimal point divisions of a section number. Example:

```
2. GENERAL INFORMATION
   2.1   Location of Equipment
   2.2   Equipment Selection Criteria
   2.3
   2.4
```

7.4.2   Design and Performance Sections for a Particular Contract

Section and subsection headings for a one-of-a-kind specification are shown in Figure 7–4 (page 7–13). Before preparing this specification, the writer should obtain the following:

- Complete information regarding the intended service of the equipment
- From equipment suppliers, general design and construction details of the equipment required

The scope of the specification should be presented first. For clarity's sake, it is desirable to keep the scope section as brief as possible and outline the technical information and requirements in the subsequent sections.

XYZ CORPORATION	ENGINEERING SPECIFICATION *(Title of Equipment)*			Sheet of
	Spec No	Contract No		Revision No

FIGURE 7–3   Example of a specification continuation sheet.

1. SCOPE

2. GENERAL INFORMATION
   [Example List]
   2.1 Reference Documents
   2.2 Equipment Location
   2.3 Miscellaneous Information
       e.g., equipment selection criteria
   2.4 Conflicting Requirements
   2.5 Language and Units of Measurement

3. RATING, CONDITIONS OF SERVICE, AND PERFORMANCE WARRANTY
   [List]

4. WORK INCLUDED
   [List]

5. WORK EXCLUDED
   [Example List]
   5.1 Installation
   5.2 Earthworks, Foundations, and Other Supports
   5.3 Architectural and Building Work
   5.4 Piping beyond Contractor's Termination Points
   5.5 Electric Power and Controls at Termination Points

6. CONTRACTOR'S TERMINATION POINTS OF SUPPLY
   [Example List]
   6.1 At Purchaser's Piping Connections
   6.2 At Equipment Supports
   6.3 At Purchaser's Connections for Power and Controls

7. APPLICABLE CODES, STANDARDS AND REGULATIONS

8. DESIGN AND CONSTRUCTION

9. DRIVER AND DRIVE ARRANGEMNETS

10. ACCESSORIES

11. INSTRUMENTATION AND CONTROL EQUIPMENT

12. SHOP INSPECTION AND TESTS
    12.1 Quality Assurance Inspection
    12.2 Inspection and Testing Requirements
    12.3 Certificates [code, material, others]

13. PERFORMANCE TESTING AT SITE
    [if specified]

14. SPARE PARTS
    [specific requirements as part of the contract]

15. EQUIPMENT WARRANTY, or DEFECTS LIABILITY
    • Months, years, or
    • Hours of operation, or
    • Minimum tonnes production, or material handled

16. CONTRACTOR'S ADVISOR(S)
    • Installation and Commissioning
    • Training Personnel

FIGURE 7–4   Sample headings for an equipment design and performance specification.

Typical examples of specification section and subsection contents are shown below.

1. SCOPE

This section is a brief statement of the work to be covered in this specification. Example:

> This specification details the general requirements for the design, construction, inspection, testing, performance and warranty of .... (*name of equipment*)

2. GENERAL INFORMATION

This section includes subsections covering general information. Examples include:

> **2.1 Reference Documents**—equipment, driver, accessories, data sheets, drawings and similar other information
>
> **2.2 Equipment Location**—whether the equipment will be installed indoors or outdoor, and be subject to a dusty, windy, seaboard, or other corrosive environment (*state which*).

Other subsections may be worded as follows:

> **2.3 Equipment Selection Criteria**
>
> The equipment furnished under this specification shall be essentially the standard product of the manufacturer. Where two or more units of the same class of equipment are furnished, these units shall be the product of a single manufacturer and full responsibility for design, manufacture, delivery date, and operation of this equipment shall be borne by the Contractor.
>
> **2.4 Conflicting Requirements**
>
> Should the Contractor's interpretation suggest a conflict between this specification, equipment data sheets, referenced standard specifications, or any other contract document, the Contractor shall seek clarification from the purchaser before proceeding with any work.
>
> **2.5 Language and Units of Measurement**
>
> All Contractor's supplied documentation including nameplates, drawings, installation instructions, operating and maintenance manuals, shall have wording in the English language.[1] All units of measurement, for example, mass, flow rate, temperature,

---

[1] or in the language of the Purchaser's country

pressure, and time, shall be in the International System of Units (SI) in accordance with .... (Standard).

3. RATING, CONDITIONS OF SERVICE, AND PERFORMANCE WARRANTY

This section details the design, performance requirements, and performance warranty of the equipment. In most instances, performance data will be detailed in the equipment data sheet(s). Examples of equipment rating and conditions of service are shown below:

3.1 The two pumps shall be identical.

3.2 The pump shut-off head shall not be more than 20 percent above the rated head specified in the pump data sheet. The pumps shall be capable of operating in parallel each taking equal load from minimum flow to the maximum without hunting, pulsating, surging, or other undesirable effects.

3.3 ....

For mining or material handling contracts, the performance requirements for a conveyor and crusher system may read:

The equipment shall be capable of conveying and crushing continuously 2,500 tonnes per hour of coal and shall be capable of handling peak loads of 2,800 tonnes per hour for periods up to 30 minutes. The conveyor shall be capable of starting from rest when fully loaded with the maximum permissible quantity of coal for up to four (4) starts per hour.

For a performance warranty, the paragraph may read:

The Contractor shall warrant that the conveying and crushing system shall be capable of a minimum continuous conveying and crushing rate of 2,500 tonnes per hour as specified in paragraph .... (relating also to the specified discharge size of coal)

4. WORK INCLUDED

This section is a detailed statement of the work the contractor will be required to do. Example:

The work shall include the furnishing of all supervision, labor, tools, equipment, materials, and drawings, and the performance of all operations and incidentals required for the construction, furnishing, and delivery of .... (*name of equipment*) complete with fittings and accessories as described herein. The work shall include:

(a) Equipment to be furnished (e.g., pumps with drivers as specified)
(b) Accessories as specified
(c) Access platforms where applicable
(d) Spare parts as specified
(e) Services of Contractor's engineer(s) for:
 • Installation and commissioning
 • Training of operators and maintenance personnel

5. WORK EXCLUDED

This section is a detailed statement of items of work which the contractor will not be required to do. A typical list is shown in Section 5 of Figure 7–4, Section 5.

6. CONTRACTOR'S TERMINATION POINTS OF SUPPLY

This section is to avoid any misunderstanding about where the contractor's scope of work terminates. An example of this item is shown in Figure 7–4, Section 6.

7. APPLICABLE CODES, STANDARDS AND REGULATIONS

This section specifies that in addition to compliance with the engineering specification(s), the design, construction, and testing of the equipment shall in all respects comply with the applicable codes and standards at the time bids were submitted. A typical list of applicable codes and standards in the United States are as follows:

NEMA - Electric motors
ASME - Boilers, pressure vessels, piping, and other major mechanical equipment
ASTM - Materials
ANSI - Flanges and piping
AGMA - Gears
OSHA - Occupational Safety and Health Administration
UL - Underwriters' Laboratories, Inc.

The equipment to be furnished shall also meet all requirements of federal, state, provincial, and local laws and regulations and other authorities having jurisdiction over the work.

8. DESIGN AND CONSTRUCTION

This section outlines paragraph titles for a centrifugal pump specification. The paragraphs are exemplary only and can be added to or deleted from as the requirements dictates.

8.1 General Arrangement

8.2 Materials of Construction

8.3 Casing

8.4 Trim

8.5 Bearings

8.6 Lubrication

8.7 Piping

8.8 Couplings and Gear

8.9 Mounting of the Equipment

8.10 .... (other paragraphs to be added as required)

9.   DRIVER AND DRIVE ARRANGEMENTS

This section covers responsibilities for furnishing, selection, and coordinating driver or drive arrangements and applicable engineering requirements, including couplings.

10.   ACCESSORIES

This section will only be included if accessories are to be supplied. Accessories to be furnished are sometimes listed in the equipment data sheet(s). The specification can spell out more explicit requirements if there is inadequate space in the data sheet(s). For example:

    (a) Battery charger for diesel engine
    (b) Free standing 500 L diesel fuel storage tank
    (c) Minimum flow orifice for each pump
    (d) ....

11.   INSTRUMENTATION AND CONTROL EQUIPMENT

This section covers the purchaser's requirements including local and remote instrumentation, the control system, control panel with mounted indicators, alarms, and safety and emergency shut-down devices.

12. SHOP INSPECTION AND TESTS

This section covers the purchaser's requirements for inspection and tests in the manufacturer's shop. The right for the purchaser to have access to the contractor's or his subcontractor's shops at any time to inspect work performed under the contract is a contract condition that has already been discussed in Chapter 5.

The scope of inspections and tests to be performed in the manufacturer's shop must be specified by the purchaser, preferably in the equipment data sheet. Some, if not all, of the inspections and tests listed may already form part of the contractor's normal quality control procedures. However, it is important that purchaser's requirements clearly state which inspection and/or test is to be witnessed or if a certified report is to be issued. A witnessed inspection and/or test has to be prearranged with the contractor, giving sufficient notice in writing for the purchaser to be in attendance.

Any additional inspection and/or test requested by the purchaser during the course of the contract that is not listed in the specification may cause a valid price extra claim by the contractor.

Also included in this section are the purchaser's quality control requirements.

13. PERFORMANCE TESTING AT SITE

This section should clearly state by whom the performance tests are to be performed. The following is a checklist of provisions to be specified:

- Test code to be used for testing
- At what stage of the contract is the test to be performed?
- Who will perform the test, purchaser or contractor?
- Who will control or supervise the test?
- Who will supply the test fluids, or other materials, and provide utility services?
- Who will calibrate and supply the testing equipment?
- Who will write the test procedure?
- Test procedure to include:
    - Sequence of tests (if there are several tests to be made)
    - Levels of performance, that is, full/partial/overload
    - Number of tests and duration of each test
    - Duration of "run-up" period
    - What readings are to be taken, where, at what intervals, and by whom?
    - Various recording sheets to be issued
    - Permissible operating adjustments during test, if any
    - Permissible running repairs during test, if any
    - Reason(s) for forced stopping of the test
    - Witnessing of recording sheets on completion of test
- Who will prepare the test report including calculations?

14. SPARE PARTS

This section specifies the purchaser's essential spare parts requirements that are to be furnished under the contract. For example, the spare parts to be supplied shall be based on 9,000 hours of equipment operation.

15. EQUIPMENT WARRANTY—DEFECTS LIABILITY

Care must be exercised to distinguish between the general warranty provisions contained in the contract conditions and those that are of technical significance and are incorporated in the technical specification. It should also be noted that the equipment warranty or defects liability (faulty design, materials, and/or workmanship) may not always specify a time period but can be in terms of a minimum production capacity. Such warranty is normally associated with material handling and mining equipment where the design criteria are based on component wear.

The equipment warranty or the defects liability period may call for an extended warranty period from the frequently accepted 12 months as purchasers seek greater operational reliability. Such provision may necessitate a change in the design including the selection of superior quality material, possibly a more sophisticated fabrication method and additional quality control procedures.

An equipment warranty for a steam and power plant may read:

> The equipment warranty period shall be 9,000 hours of operation for each boiler and all its auxiliary equipment, and each back-pressure turbine generator and all its auxiliary equipment, provided that such warranty period does not exceed 24 months from the date of equipment take over. For all other equipment, the warranty period shall be 24 months.
>
> The aforesaid hours of operation shall not be deemed to have commenced until the equipment has been in continuous operation for a period of 28 days under continuous load without fault as specified in paragraph ....

16. CONTRACTOR'S SERVICES AT SITE

This section should clearly specify the purchaser's requirements as part of the contract for:

- Assistance with installation of the equipment
- Assistance with commissioning of the equipment
- Training operator and maintenance personnel

Emphasis should be placed on the following:

- Contractor's engineer(s) are to be fluent in speaking English (or of language at site)
- Duration of time (days) for personnel training to be specified
- Contractor should provide a proposed training program for the purchaser's review and acceptance

7.4.3   General Data and Requirements Sections Applicable to Most Contracts

The following general data and requirement sections are usually standard for each contract. A list of sample section and subsection headings for the general requirements are shown in Figure 7–5 (page 7–21).

1. TO 3.   GENERAL DATA

The information shown in Sections 1 to 3 should be completed where applicable for the particular contract.

4.   MATERIALS AND WORKMANSHIP

This section is a standard specification requirement and normally reads as follows:

> Materials used for work performed under the contract shall be new, of the highest quality, and suitable for the intended service conditions of the equipment

> Work shall be performed by skilled mechanics in the various crafts and be executed in a first class and workmanlike manner in accordance with recognized good practice in the industry

5.   DRAWINGS AND TECHNICAL DATA SUBMITTAL REQUIREMENTS

The scope of drawings and technical data to be submitted for the purchaser's review and the submission periods after the award of contract are to be specified by the purchaser in the *Drawing and Technical Data Submittal Requirements Form*, details of which are discussed in Paragraph 7.6.2. This form also lists the drawings each bidder has to submit with the bid. It should be clearly understood that the drawings and other technical data to be furnished by the contractor form part of the contract.

The drawings and technical data to be submitted after award of the contract serve the following three purposes:

1. To check that the contractor fully understands the purchaser's requirements
2. To illustrate and detail the purchaser/contractor termination points of supply
3. To provide information the purchaser needs to know for engineering work by others, such as mass of components, layout details for foundations, and installation requirements

6.   INSTALLATION INSTRUCTIONS, OPERATING AND MAINTENANCE MANUALS

Experience has shown that inadequate attention is given to outlining the purchaser's precise requirements for the scope and content of the operating and maintenance manuals and the number of sets required. The costs to compile these documents especially for complex

GENERAL DATA

1. SITE CONDITIONS
    - Ambient temperature
        - minimum / maximum / average / design
    - Altitude – above / below sea level
    - Atmospheric pressure – normal yearly
        average
    - Relative humidity – average / maximum
    - Design wind category
    - Seismic factor
    - Environmental conditions
        - inland / seaboard / dusty / corrosive
    - Rainfall
        - Annual average
        - Maximum intensity

2. UTILITIES AVAILABLE
    2.1 Power Supply
    Voltage / Phase / Frequency
        - Motors
        - Lighting
        - General purpose power
        - Instrumentation and controls
    2.2 Compressed Air
        - Instrument (oil free) pressure
        - Service (utility) air pressure
    2.3 Water
        - Process water (non drinking)
            pressure and temperature
        - Potable water
            pressure and temperature
        - Quality of process water for design
    2.4 Steam
        - Supply pressure and temperature

3. EQUIPMENT NOISE LIMITATION
    - Terminology
    - Measuring device and method
    - Limitation values

GENERAL REQUIREMENTS

4. MATERIALS AND WORKMANSHIP

5. DRAWING AND TECHNICAL DATA
    SUBMITTAL REQUIREMENTS

6. INSTALLATION INSTRUCTIONS,
    OPERATING AND MAINTENANCE
    MANUALS

7. NAMEPLATES, MARKINGS, AND
    NOTICES

8. CLEANING, PAINTING, AND
    PRESERVATION

9. SPECIAL TOOLS AND DEVICES

10. ASSEMBLY, PACKAGING, AND
    SHIPPING

11. SPARE PARTS [as a standard
    specification requirement]
    - Construction
    - Maintenance
    - Consumables

12. WORK PROGRAM

13. PROGRESS CONTROL AND
    REPORTING

FIGURE 7–5   Sample headings for a general data and requirements specification.

equipment can be extremely high. Unless the purchaser specifies the complete requirements, contractors in general are reluctant to furnish more than the basic information.

In many instances, this is a collection of preprinted "off-the-shelf" brochures and very often not applicable to the actual equipment supplied. Furthermore if the purchaser does not specify the number of copies required in the bidding documents, bidders will then quote a bare minimum number of copies. An oversight on the number of copies required and detail of contents can prove extremely costly to the purchaser when making a request for this information to be provided after award of the contract.

Specified requirements should include:

- Each set of manuals to be divided into separate bound volumes for:
  - Installation instructions
  - Operating instructions
  - Maintenance instructions with list of spare parts
- Number of bound sets of manuals to be submitted, and when required by the purchaser, as noted in the drawing and data technical requirements form
- All information and drawings should be furnished for binding in 216 mm by 279 mm binders unless specified otherwise
- Maximum thickness of each volume to be specified
- Type of binding

Rather than duplicate here a lengthy list of items and information to be included in the manuals, refer to Part 5, Section 15.2.5, General Data and Requirements Specification, for an example of a major equipment procurement.

7.   NAMEPLATES, MARKINGS, AND NOTICES

All major equipment, whether required by statutory regulations or otherwise, should have suitable nameplates affixed and appropriate marking prior to shipment. Nameplate material (especially in corrosive environments), location, and method of fixing should be specified, rather than leaving this to the contractor's discretion. Nameplates must still be visible where equipment has been insulated.

Items imprinted on nameplates include:

- Purchaser's contract number
- Equipment tag number
- Manufacturer's name and country of manufacture
- Manufacturer's information—type, serial number, date of manufacture, etc.
- Performance data—capacity, pressure, temperature, speed, power, etc.
- Information required in accordance with statutory regulations

Suitable nameplates shall also be attached to valves and gauges furnished with the equipment.

Equipment markings should include:

- Arrow for direction of rotation
- Direction of flow in piping

Notices to be provided should include:

- Location(s) for lifting
- For hoist, monorails, and cranes, the rated capacity, such as the maximum allowable lifting mass in kilograms
- Warning signs, such as "Danger"

8. CLEANING, PAINTING AND PRESERVATION

Many organizations have a standard engineering specification covering cleaning, painting and preservation of the work. The following suggested language may be used as a check-list of general requirements.

8.1 Cleaning

Equipment shall be neatly finished all over to give a smooth appearance. All exposed or external surfaces shall be free from burrs, fins, mill scale, oil, and other foreign matter. Castings must be sound, free from blow-holes, cold shuts, sinks, or tool marks. Patching, peening, or caulking is not permitted.
Plates, structural steel members, and other steel components shall be sand or grit blasted to a standard finish. In some cases, sand blasting is not permissible. Also acid dipping is sometimes specified in lieu of blasting.

8.2 Painting

External surfaces that are not machined and that have been thoroughly cleaned shall be painted within four hours with the manufacturer's standard metal primer followed by at least two top coats of the manufacturer's standard finish paint, unless otherwise specified.

8.3 Preservation

Machined surfaces of the equipment shall be coated with a suitable rust preventative to prevent deterioration prior to installation. Other surfaces subject to corrosion shall be coated with a corrosive preventative that is readily removable with a commercial solvent at the site prior to start-up of the equipment.

9. SPECIAL TOOLS AND DEVICES

It is of major importance that the contractor furnishes as part of the contract all of the special tools and devices the purchaser will need for operating and maintaining the equipment supplied. The use of these should be explained in the maintenance manuals.

All such tools and devices should be new, shipped separately from the main equipment, and be appropriately marked with the name of the tool to identify its intended use.

10. ASSEMBLY, PACKAGING, AND SHIPPING

Packing and shipping are normally the contractor's responsibilities unless specifically stated otherwise in the contract conditions. Adequacy of crating, handling, and environmental protection of the equipment while in transit has frequently been given insufficient attention by contractors. The equipment should be strongly packed and crated in accordance with the best commercial practice. Rectification of consequential damage is the contractor's responsibility and if severe, can cause a delay in the completion of the procurement.

Before shipment, all openings of the equipment should be protected by means of metal or wooden blank flanges (whichever is specified) fully bolted in place if flanged, metal plugs if screwed, or metal caps if weld end.

The equipment and all components should preferably be shipped to the site in a completely assembled condition or as complete as possible within shipping and handling limitations. For driver and driven equipment, it is recommended, that couplings be disengaged, and where gears are furnished, that they be adequately protected against rocking motion during shipment to avoid brinelling of bearings, gears, and shafts.

Each package should be clearly marked with the purchaser's contract and item number so as to facilitate identification when received at the site.

Not always recognized when preparing the equipment for shipment is that the contractor should be fully aware of the quarantine regulations at the destination port. Straw packaging is often prohibited. If this has been overlooked, fumigation may be required by the port authority at the contractor's expense. Also, timber packing or casing requires fumigation for import into many countries.

11. SPARE PARTS

This section deals with spare parts to be supplied by the contractor as part of the contract (see Paragraph 7.4.2, Section 14) and recommended optional spares.

Spare parts are divided in the following categories:

● Selected spares required for commissioning and possibly performance test purposes. The supply of these spares form part of the contract.

- Spare parts to be kept as an inventory in the purchaser's store for equipment maintenance as recommended by the contractor. This list of spare parts together with unit prices is to be submitted with the bid. The list is accepted or amended by the purchaser during negotiation with the successful bidder. The final negotiated list becomes part of the contract.

The purchaser's requirements must be specified in the specification including the period of time, for which the spares are required, for example, after 9,000 hours of operation.

All spare parts are to be listed in the maintenance manuals.

12. PROGRESS REPORTING

This section generally applies to long delivery procurements. It should be noted that in some cases progress reporting is a contract condition rather than a technical requirement. In any case, the provision should not be duplicated.

As part of the scope of the work, the contractor should submit progress reports at specified intervals (usually once a month) to cover the following aspects of the work:

- Status of the work in progress with current projected completion dates
- Work completed since the previous report including dates of completion
- Work rescheduled from the contract program dates together with reasons therefore
- Work where current or anticipated delays will affect contract completion dates together with remedial actions to be taken by the contractor

Such matters as milestone date changes to the contract program and action to be taken by the purchaser should work fall behind schedule because of the contractor's remedial measures proving to be unsatisfactory are commercial issues and are covered in the contract conditions.

7.4.4   Checking and Revising the Specification

Checking the Specification

The following items are to be again verified just prior to issuing the specification for bidding and for purchase:

- Clearly defined the scope of work to be performed
- Technical requirements in conformance with the project design criteria
- Proper and complete reference to data sheets and drawings
- Compliance with current standards, codes, and regulations

- Coordination of requirements specified in other specifications applicable to the contract
- Inclusion and validity and currency of applicable engineering specifications, standards, codes, and regulations

Revising the Specification

It is important that a specification be maintained at a current status at all times and that those who are involved with the contract, receive a revised copy each time a revision has been made to the document.

Recommended procedures for revising a specification are:

Revisions to a section of a paragraph within the body of the specification are made by lining out the portion to be revised and neatly printing the new information above it. A revision is noted by adding a revision triangle opposite the paragraph being revised with the revision number flagged in the triangle ($\triangle$ for Revision 1). The latest revision number is to be added to the block on the specification page. The "Issue Description" block on Sheet 1 should note "Paragraph—revised as noted."

Deletions are made in one of two methods: retyping the page, or by lining out the portion to be deleted. In each case, a revision triangle must be inserted. For retyping, the word "deleted" should be written (or typed) in. The preferred method is by lining out, as the revised copy will always show what information has been deleted without having to make reference to the previous issue. It also saves time and effort, providing it is carried out without loss of neatness or clarity. In no case should a paragraph number be removed, nor must paragraphs be renumbered from the original. Further, a new or additional paragraph should not be inserted under a deleted paragraph.

Additions requiring paragraphs to be enlarged or where new paragraphs are to be added, can be handled as follows:

1. Where numerous major changes are required throughout the specification, retyping is generally required, or
2. Where, on one page, a paragraph is to be enlarged, or additional paragraphs added, the revision is done by retyping the page with the corrections and continuing the remaining paragraphs on a new sheet, for example, Sheet 4A. The total number of pages shown on Sheet 1 remains the same (say 12), but "plus Sheet 4A" is shown typed in above "Sheet 1 of 12." The "Issue Description" block should note "Added Sheet 4A." The revision triangle, flagged with the revision number, must be inserted where the addition was made.

7.5   EQUIPMENT DATA SHEETS

7.5.1   General

Experience has shown that bidders do not always submit the technical information the purchaser is expecting to receive for a major equipment procurement. This may not be the bidder's fault. A lack of clearly defined requirements in the bidding documents will leave bidders little option other than to assume what should be included in their bids. Manufacturer's brochures showing only general "boilerplate" data are frequently submitted, and in many instances, the information is not related to the particular equipment being solicited. Furthermore, bidders are reluctant to commit large amounts of time and money for bid preparation. The information then received by the purchaser is often insufficient to fully assess the bid.

These situations can be avoided by the use of an *Equipment Data Sheet*. An equipment data sheet is a standard engineering fill-in sheet where certain portions are completed by the purchaser to specify the technical requirements before releasing the bidding documents. The remainder of the fill-in portions are completed by the bidders relative to the equipment offered. As discussed in previous sections, engineering specifications can cover a standard set of technical requirements, whereas the data sheets supplement the specifications by specifying the operating, performance, construction, and other features of the equipment to be supplied for a particular procurement. Should conflict of information arise between an engineering specification and a data sheet, the information in the data sheet should take precedence.

The equipment data sheet serves the following functions:

- All bidders complete the same fill-in portions of the sheet. This minimizes the risk of incorrect information being submitted.
- They are very effective checklists of the technical features of the equipment to be supplied.
- They save bidder's time in the process of selecting the correct equipment for the intended service.
- They avoid unnecessary queries and correspondence with bidders concerning the details of equipment offered.
- When all fill-in portions are completed by both the purchaser and bidders, the data sheet should contain virtually all of the essential technical information necessary for evaluating the bids.
- After procurement, the data sheet becomes a record of the major technical information of the equipment that has been supplied. Reference can be made to the document to check the equipment's performance capabilities should process upgrading be required or the equipment be relocated for another service.

Data sheets can be used for items of equipment such as:

- Mechanical:
  Rotating equipment: pumps, compressors, fans, engines, turbines and auxiliary equipment including gears and lubrication systems
  Heat exchangers: shell and tube, air cooled, cooling towers, fired heaters, including steam generators
  Tanks and vessels: storage tanks, pressure vessels, vacuum vessels
- Electrical:
  Motors, switchgear, transformers
- Instrumentation and controls:
  Items for local and remote services, alarms
- Material handling:
  Belt conveyors, trippers, bucket elevators, lifting magnets, weighers, cranes, and hoists
- Mineral processing:
  Feeders, mills, crushers, separators, classifiers, agitators and mixers, thickeners, flotation, sampling equipment

Equipment data sheets are not only developed by individual companies and government agencies but also form an integral part of some U.S. standards, including those published by the American Petroleum Institute (API), Tubular Exchanger Manufacturers Association (TEMA), and the Instrument Society of America (ISA).

A data sheet has a document title, document number, and revision number and hence is an attachment noted in the technical requirements form. It is treated and revised in the same manner as a drawing. Whereas equipment operating and performance data are normally the only items requiring revision, there should be no need to rework any part of what could be a lengthy specification.

It should be recognized that a data sheet is an engineering document and should not contain information on price, delivery, or other commercial matters.

7.5.2   Typical Framework of the Data Sheet

Discussions on the formation and content of data sheets for electrical equipment and instrumentation are for the specialists working in those sections of industry and will not be considered in this book except for electrical motors driving mechanical equipment and machine-mounted instrument packages. The absence of discussion about these data sheets is not intended to minimize their importance for major equipment procurements, but rather to limit the scope of this book. For easier understanding, examples of a equipment data sheets are shown in Figures 7–6 (page 7–29), 7–7 (page 7–30), and 7–8 (page 7–31).

1	Project		Plant			Service			
2	Manufacturer		Size / Type			Serial No.			
3	Type Of Driver		Furnished By			Mounted By			
4	Pump Specification No.			Pump General Arrangement Dwg No					
5	Other Applicable Specifications								
6		OPERATING CONDITIONS					PERFORMANCE		
7	Liquid Pumped					Performance Curve No.			
8	Capacity Flow	Normal	Rated		L / s	Speed	rpm	NPSH Reqd	m
9	Liquid Temp	Normal	Rated		°C	Efficiency	%	kW Rated	
10	Discharge Press	Normal	Rated		kPag	Max kW Rated Impeller			
11	Suction Press	Max	Rated		kPag	Max Head Rated Impeller			m
12	Differential Press		kPa		m	Rotation: from Coupling End – cw / ccw			
13	Specific Gravity at Rated Conditions					SHOP INSPECTION AND TESTS			
14	NPSH Available			m		"x" where applicable	Certified	Witnessed	
15	Vapor Press at Rated Conditions			kPaA		Perform w/o Driver			
16	Viscosity at Rated Conditions			Pa • s		Perform with Driver			
17	Corrosion / Erosion Caused By					Hydrostatic Test			
18		CONSTRUCTION				NPSH Test			
19	Nozzles	Size	Rating	Facing	Location	Shop Inspection			
20	Suction					Dismantle & Inspect			
21	Discharge					After Test			
22	Case – Mounting Centerline / Foot / Bracket								
23	Split Axial / Radial						MATERIALS		
23	Type Volute Single / Double / Diffuser					Pump Case			
24	Hydrostatic Pressure		kPag			Case Wear Rings			
25	Connection Vent / Drain / Gage					Impeller			
26	Impeller Diameter Rated	/ Max	/ Min		mm	Impeller Wear Rings			
27	Mounted Between Bearings / Overhung					Shaft			
28	Bearings – Type Radial		Thrust			Sleeve (Packed)			
29	Lubrication Ring Oil / Flood / Flinger / Pressure / Grease					Sleeve (Seal)			
30	Coupling Mfr	Model				Throat Bushing			
31	Driver Half Mounted By: Pump Mfr / Driver Mfr / Purchaser					Gland			
32	Packing Type		No. of Rings			Lantern Ring			
33	Mechanical Seal Mfr & Model / Type					Base Plate			
34	Mfr Std No.					MASS OF EQUIPMENT			
35	Base Plate: Common with Driver yes/no; Drip Rim yes/no					Pump Net	/ Pump & Driver		kg
36		DRIVER							
37		Electric Motor	Steam Turbine			Engine		Gear	
38	Manufacturer								
39	kW / rpm								
40	Spec No.								
41	Data Sheet No.								
R									
E									
V									
	No.	Date		ISSUE DESCRIPTION			By	App'd	Date

XYZ CORPORATION	HORIZONTAL CENTRIFUGAL PUMP EQUIPMENT DATA SHEET	Document No.
	Equipment Tag No. Contract No.	Revision No.

FIGURE 7–6   Centrifugal pump data sheet.

1	Project	Plant	Service	
2	Manufacturer	Size / Type	Serial No.	
3	OPERATING CONDITIONS			
4	Driven Machine		Fuel Type	
5	Type of Service		Gas Fuel Low H V	kJ / m³
6	Power Required	kW	Supply Pressure	kPag
7	Speed Required	rpm	Liquid Fuel Low H V	kJ / m³
8	Altitude	m	Viscosity @ ° C	Pa • s
9	Ambient Temperature ° C		API Gravity @ Deg	
10	Outdoors / Indoors / Roofed / Curtain Wall		Cetane or Octane Rating	
11	DESIGN AND PERFORMANCE			
12	Arrangement		Turbocharged	Compression Ratio
13	No. of Cylinders: / Bore / Stroke mm		4 or 2 Cycle Displacement m m ³	
14	Cooling Water Temp. ° C In / Out		RATINGS	
15	Engine Cylinders /		This Service / Rated / Max Cont	
17	Turbocharger /		kW	/ /
18	Aftercooler /		Speed rpm	/ /
19	Oil Cooler /		kJ / kW / h	/ /
20	Exhaust Manifold /		Piston Speed	/ /
21	MATERIALS OF CONSTRUCTION			
22	Frame	Bearings:	Valves:	
23	Crank Shaft	Main, No. & Size	Intake, No. & Size	
24	Cylinders	Material	Material	
25	Liners	Crankpin, Size	Seat Material	
26	Pistons	Material	Exhaust, No. & Size	
27	Cylinder Heads	Wristpin, Size	Material	
28		Material	Seat Material	
29	EQUIPMENT AND ACCESSORIES TO BE FURNISHED BY THE SUPPLIER			
30	Turbocharger	Exhaust Manifold	Pyrometer and Selector Switch	
31	Turbo Air Aftercooler	Exhaust Silencer	Tachometer	
32	Ignition System	Radiator and Fan	High Water Temperature Shutdown	
33	Magneto or Pulse Generator	J W Heater Exchanger	Low Oil Pressure Shutdown	
34	Governor	J W Pump	Engine Overspeed Shutdown	
35	Air Head for Governor	Lube Oil Pump / Cooler	Turbo Overspeed Shutdown	
36	Throttle Control	Lube Oil Filter Single / Dual	Thermometers – Water	
37	Fuel Gas Regulator & Shut–off	Pre – Lube Pump	Thermometers – Oil System	
38	Fuel Gas Volume Bottle	Engine Lubricator	Thermometers – Exhaust System	
39	Fuel Pump / Fuel Filter	Oil Cooled Pistons	Flywheel and Guard	
40	Fuel Day Tank	Batteries	Flywheel Barring Device	
41	Starting Air Valves and Piping	Clutch and Power Take–off	All Weather Sheet Metal Housing	
42	Start Motor Air or Electric	Gauge Board		
43	Air Intake Filter	Operating Panel		
44	Overall Dimensions Length mm, Width mm Lifting Mass kg			
R				
E				
V	No. Date	ISSUE DESCRIPTION	By	App'd	Date

XYZ CORPORATION	**INTERNAL COMBUSTION ENGINE EQUIPMENT DATA SHEET**	Document No.
	Equipment Tag No. Contract No.	Revision No.

FIGURE 7–7   Internal combustion engine data sheet.

1	SITE CONDITIONS
2	Location Environment: Inland / Seaboard / Dusty and Windy / Corrosive / Other
3	Ambient Temperature: Max / Min / Design ° C Altitude m
4	BASIC DATA AND REQUIREMENTS

5	Purchaser's Motor Specification No		
6	Manufacturer		
7	Frame Designation		
8	Type		
9	Rating Output (kW)		
10	Rated Voltage		
11	Phases / Frequency		
12	Service Factor		
13	Type of Construction		
14	Method of Cooling		
15	Protection		
17	Rated Current (amps) FL		
18	Rated Speed (rpm) Syn / FL		View on End Drive
19	Rotation – Driving End		
20	Efficiency: 100%, 75%, 50% FL		
21	Power Factor: 100%, 75%, 50% FL		
22	Locked Rotor Current		
23	Rated Torque (Nm)		
24	Torque: Starting / Pull Out (% FLT)		
25	Inertia of Rotor		
26	Design Standard / Design Letter		
27	Enclosure		Side View
28	Insulation Class		
29	Time Rating / Temp Rise ° C		
30	Winding Connection		
31	Bearing Type / Make		Location of Terminal Box and
32	Type of Lubrication		Entry: _____
33	Slide Rails / Flange / Size		
34	Heaters: No. / Voltage / Rating		
35	Temperature Detect. Type / No. / Temp		
36	Allowable Number of Starts / Hour		
37	Motor Mass (kg)		
38	Mounting Drawing No		
39	Termination Drawing No		

40	Noise – Sound Pressure Level at 1 m @ Full Load dB(A)
41	Motor Performance Curve No: General Arrangement Dwg No:
42	Witnessed Inspection and Testing: With Driven Equipment Yes / No
43	Furnished Options: Refer to Specification

R						
E						
V						
	No.	Date	ISSUE DESCRIPTION	By	App'd	Date

XYZ CORPORATION	**HORIZONTAL INDUCTION MOTOR DATA SHEET** Document No. Driven Equipment Tag No. Revision No. Contract No.

FIGURE 7–8   Induction motor data sheet.

The standard data sheet form is normally divided in the following sections with fill-in information to be completed by both the purchaser and the bidder as the case may be:

- Name of plant, area location, equipment service—by the purchaser
- Name of manufacturer, country of manufacture, equipment type, serial number, and other identification—by the bidder
- Responsibility for furnishing the driver and couplings—nominated by the purchaser
- Operating conditions—by the purchaser
- Performance information of the equipment offered—by the bidder
- Construction—by the bidder except for certain items specified by the purchaser such as:
  - Equipment nozzle ratings
  - Mechanical seal or gland packing
  - Common base plate required or otherwise

  Should the driver be furnished by the purchaser, it is important to specify who will supply and install the coupling as well as who is responsible for alignment of driver and driven equipment. Also, who will mount the driver—nominated by purchaser
- Materials—by bidder unless the purchaser specifies a component's material
- Scope of shop inspections and tests—nominated by the purchaser.

  Purchaser to indicate which tests are to be witnessed and those that are not witnessed but where a certified report is required. Because of the importance of this part of the data sheet, the following is a representative list of inspections and tests:

Shop inspection	Auxiliary equipment run
Hydrostatic	Dismantle, inspect and reassemble
NPSH	Performance test
Mechanical run with driver	Special test(s), e.g., overspeed
Mechanical run without driver	Vibration
Run spare rotor	Noise

- Summarized information for the equipment driver if furnished by the contractor—nominated by the bidder. A separate data sheet is usually issued by the purchaser for the driver
- Accessory equipment noted in applicable specifications, to be furnished by the contractor—nominated by the purchaser
- Major equipment dimensions and equipment mass—by the bidder
- Equipment performance specification—nominated by the purchaser
- Other applicable specifications, such as noise level, painting—nominated by the purchaser
- Applicable reference drawings—nominated by the purchaser

7.6   DRAWINGS

Drawings are one of most effective ways for transmitting graphical information both from the purchaser to the contractor and from the contractor to the purchaser. This section discusses the following:

- Drawings issued by the purchaser to supplement the engineering specifications
- Drawings and technical data to be submitted by the contractor for the purchaser's review and those necessary for the purchaser to complete the engineering of the project based on the contractor's information

7.6.1   Drawings Issued by the Purchaser

Drawings issued by the purchaser are usually for reference purposes to supplement the engineering specification(s). These drawings, together with their latest revision number, are listed as attachments in the technical data requirements form, and normally include:

Site data—site and equipment location. Rail and road routes and clearances for vehicles for delivery of the equipment to site are the contractor's responsibility.

General information drawings—to be used by the contractor for reference purposes and which include purchaser's piping and instrument diagrams, location of termination points of supply, electrical single line diagrams (if appropriate), and a plot plan that fully coordinates the location of all major items of equipment.

Company standard engineering drawings—to be used whenever applicable in the contractor's design and include such details as: platforms, stairways, ladders, and handrails; pipe fittings and supports; footing details; vessel supports, vessel nozzles, connection details to piping and vessels; electric cabling details; instrumentation and control information.

The following should be noted with regard to drawings listed as attachments in the technical requirements form:

- Certain drawings may be issued for bidding purposes only. Others are to be used for reference during the course of the contract. The drawings should therefore be clearly stamped whether they are issued for construction and marked "Approved for Construction," or be used for reference only and stamped "For Information Purposes Only," or "Drawing Unchecked," or "Not to be used for construction." All stamped markings on drawings are of large size print with the first letter of each word in the upper case to highlight their importance.

- Work to be performed by the contractor and work that will be performed by the purchaser should be clearly indicated on each drawing
- Drawings must have the same nomenclature as the engineering specifications. Where there is a discrepancy between a drawing and a specification, the specification normally takes precedence or as otherwise defined in the contract documents

7.6.2   Drawings and Technical Data to Be Submitted by the Contractor

Where the contractor is responsible for the design of the equipment, the technical specification should have a provision requiring the contractor to submit certain drawings and data at times specified by the purchaser. This information is also referred to as vendor data, or vendor prints, or vendor drawings. The accurate and timely provision of information shown is vital to the success of the project and its completion on time. The material when received will serve the following three purposes:

1. Ensure that the contractor and the subcontractor(s) have a complete understanding of the equipment to be supplied, and that it complies with the purchaser's technical specification.
2. Permit the purchaser to complete the engineering of the project for such items as foundations and at equipment interfaces—piping, electrical, instrumentation and controls.
3. Provide necessary information on installation, instructions, and operating and maintenance of the equipment being supplied.

The purchaser is responsible for determining, in a precise and orderly manner, what information is to be provided and when, so as to ensure that the total project will be completed on time. The sequence of submitting this information must be in conformance with the project program and be carefully planned, especially where multidiscipline engineering is involved. This can best be achieved with the use of a standard *Drawing and Technical Data Submittal Requirements Form*, shown in Figure 7–9 (page 7–35).

    The drawing and technical data submittal requirements form becomes part of the technical specification, having been listed as an attachment in the technical requirements form (see Section 7.1). Since it is included in the bidding documents, it subject to acceptance by the bidder or to negotiation during bid evaluation. The submittal of the this information means time and money to the bidder, particularly where the equipment is complex and a large amount of information is needed. It must therefore be accounted for and included in the bid estimate.

    After signing the contract agreement, the form is part of the contract and must be

XYZ CORPORATION

DRAWING AND TECHNICAL DATA SUBMITTAL REQUIREMENTS

CONTRACT No   SHEET 1 of   REVISION No   DATE

PRINTED MATERIAL AND DRAWINGS TO BE SUPPLIED BY THE CONTRACTOR AS INDICATED

TECHNICAL DATA FOR .... Equipment Tag No	PRINTS WITH BID	REVIEW CODE (1)	CERTIFIED INFORMATION AFTER AWARD			
			No. of DWGS		No. of	WHEN
			TRANS– P'ANCY	PRINTS	PRINTED MATERIAL	REQ'D (2)
1 Drawing List and Drawing Schedule						
2 General Arrangement Drawings						
3 Dimensional Outline						
4 Equipment Lifting Mass (kg)						
5 Foundation Plan / Anchor Bolts						
6 Load Diagrams – Foundations / Nozzles / Other Loadings						
7 Section and Details						
8 Performance Data						
9 Completed Equipment Data Sheets						
10 Mechanical Terminal Connection Data						
11 Electrical Schematic and Wiring Dwgs						
12 Calculations						
13 Welding Procedures and Qualifications						
14 Installation Instruction Manuals						
15 Operating Instruction Manuals						
16 Maintenance Manuals and Complete Parts List						
17 Certificate – Statutory – Material Test(s) – Performance Test(s)						
18 Spare Parts – part of the contract						
19						
20						

Legend : (1) Review Code F :   Review is required prior to fabrication
                        S :   Review is required prior to shipment
                        R1 :   One copy only to be submitted for first review
                        I :   Required for information only

          (2) In weeks after award of Contract, or "B/S" before shipment, or as specified

Unless specified otherwise, all drawings and data shall be directed to:

          Attention (name of person)   ....
                (name of company)   ....
                        (address)   ....
                                    ....

FIGURE 7–9   Example of a drawing and technical data submittal requirements form.

strictly adhered to by the contractor. At the same time, the purchaser cannot exercise his or her right to demand the contractor to submit additional information over and above what is contained in the drawing and technical data submittal requirements form or in the other engineering documents that form the technical specification for the contract.

It should be noted that the contract conditions do not always define the meaning of "approval" by the purchaser of drawings, data, and other information submitted by the contractor. It must be understood that the contractor remains responsible for the design and supply of the equipment in accordance with the conditions of the contract regardless of the purchaser's "approval" or "approval with comment" to a document submitted. It is for this reason that the term *review* should be used in preference to *approval* to avoid a dispute arising later as to the interpretation of the contractor's right in this matter.

The following information must be made available to the contractor:

- Drawing title block requirements
- Allocation of drawing numbers
- Drawing sheet sizes

Other than bid prints, certificates, and manuals, drawings should be submitted in transparency form. This will permit the purchaser to make multiple copies of the reviewed material for distribution to the contractor, purchaser's engineers, designers, inspectors, expediters, construction, and production departments.

The drawings and data listed in Figure 7–9 are illustrative only. The list would vary to suit the particular equipment being procured such as for pumps, compressors, heat exchangers, vessels, and other miscellaneous equipment. The purchaser should not request more information than is necessary. Unnecessary information increases the reviewer's workload and consequential costs as well as document handling charges (recording, distribution, filing and mailing). Asking for too little information may lead to costly consequences should incorrect or unsuitable equipment be received at site.

The engineering specification should make reference to the requirement that on completion of the contract, the contractor is to furnish the purchaser with a complete set of "as built" drawings either in transparency form or as microfilm as specified.

The following are definitions of key engineering documents normally required to be submitted:

Drawing List and Drawing Schedule. A complete list of all drawings and data that the contractor is required to supply. The contractor shall also submit a program of completion for all drawings on the list within the periods stated in the submittal requirements form.

General Arrangement Drawings. These are generally outline drawings of the equipment to scale, including dimensioned location of related accessories.

Detailed Drawings. These show all details of components required for assembly of the equipment, and must be cross-referenced to other drawings.

Foundation Plan and Anchor Bolts. These show all the necessary dimensions and details required for foundation and setting design, including location, embedded items, size, type and projection of anchor bolts.

Load Diagram. This provision provides the total static and dynamic loads and points of application.

Spare Parts. This is a list of components with part names and numbers indicated and referenced to equipment drawings, and maintenance manuals.

Revising the Drawing and Technical Data Requirements Form

The drawing and technical data submittal form is first issued with the bidding documents, and is designated with the revision status Revision 0 (or simply Rev. 0). It retains this revision status unless subsequent changes are made to the document.

Review Procedures of Contractor's Submissions

The person who determines the drawing and technical data submittal requirements should also have a clear understanding of the information reviewing procedures. Clerical formalities such as date stamping the documents when received and proper distribution before and after review, are extremely important, but are in-house procedures not discussed in this book. It should be noted however, that careful planning for handling and processing the documents is important. Since the contractor is instructed to withhold manufacture until the results of the purchaser's review is received, a delay in the dispatch of such information may delay the delivery of the equipment.

The submitted document is reviewed, marked-up where necessary, and returned to the contractor within 14 days after receipt by the purchaser. The result of the review is noted on the purchaser's review stamp affixed to the document. Examples of the review stamp are shown in Figures 7–10 and 7–11 (page 7–38). The review stamp shown in Figure 7–10 is for the submission of drawings where the purchaser's review is required prior to the commencement of fabrication.

The review stamp shown in Figure 7–11 is for the review of such documents as certificates, installation instructions, and operating and maintenance manuals. The note "Reviewed—No Comment" shall be construed to mean that the information submitted complies with the general features, and arrangement requirements. "Accepted Except Where Noted" shall be construed to mean that the exceptions noted shall be corrected by the contractor and resubmitted to the purchaser as certified.

```
+-----------------------------------------------------------+
|                   XYZ  CORPORATION                        |
| REVIEWED BY  .........................    DATE ..........  |
|-----------------------------------------------------------|
| THIS REVIEW DOES NOT RELIEVE THE CONTRACTOR'S RESPONSIBILITY|
|     FOR THE DESIGN AND COMPLIANCE WITH THE CONTRACT        |
|-----------------------------------------------------------|
|        FINAL           |        CORRECT AND RESUBMIT       |
|                        |                                  |
| RELEASED FOR FABRICATION| BY  ............................ |
|                        |                                  |
|  [ ] 1  NO COMMENT     |  [ ] 3  RELEASED FOR FABRICATION |
|                        |          EXCEPT AS NOTED          |
|  [ ] 2  WITH COMMENT   |                                  |
|                        |  [ ] 4  NOT RELEASED FOR         |
|                        |          FABRICATION             |
+-----------------------------------------------------------+
```

FIGURE 7–10   Contractor drawing review stamp.

```
+--------------------------------------------------+
|                REVIEWED AS NOTED                 |
|    BY ..............    DATE .................    |
|                                                  |
|   [ ]   REVIEWED – NO COMMENT                    |
|                                                  |
|   [ ]   ACCEPTED EXCEPT WHERE NOTED              |
|          RESUBMIT BY ..........................  |
|                                                  |
|   [ ]   NOT ACCEPTED                             |
|          RESUBMIT BY ..........................  |
|                                                  |
|--------------------------------------------------|
|  THIS REVIEW DOES NOT RELIEVE THE                |
|  CONTRACTOR'S RESPONSIBILITY FOR                 |
|  THE DESIGN AND COMPLIANCE WITH                  |
|  THE CONTRACT                                    |
|--------------------------------------------------|
|                XYZ  CORPORATION                  |
+--------------------------------------------------+
```

FIGURE 7–11   Contractor certificate and manual review stamp.

For both review stamps, the information provided shall not relieve the contractor from responsibility for errors, omissions, or for adequate design, performance, and proper operation of the equipment, and compliance with the contract.

8

DRAFTING THE
REQUIREMENTS FOR BIDDING

8.0 INTRODUCTION

The preparation of the commercial and technical documents for bidding was outlined in the previous three chapters. This chapter describes recommended guidelines for drafting the bidding requirements for inviting bids by the purchaser. It should be recognized that these procedures will vary depending on the type and scope of the procurement undertaken. However, the essential requirements should vary very little from one procurement to another.

It is recommended that procedures for bidding be set out in four documents identified as:

1. *Invitation to Bid.* A document inviting prospective bidders to submit a bid to undertake the work under the contract
2. *Instructions to Bidders.* A document setting out the instructions and conditions under which bids are to be prepared, submitted, received, and evaluated by the purchaser
3. *Content of the Bid.* A document listing of the purchaser's data and information (schedules) to be completed and submitted by bidders
4. *Bid Form.* A document completed and signed by the bidder that is legally deemed to be an offer to execute the work under the contract

It is important to understand the purpose of each of the above documents. The documents should contain only information and instructions that are relevant to the preparation and submission of bids (see Chapter 3 and Figure 3–2).

The following is a brief outline of the scope of each of the documents making up a typical set of requirements for bidding for a large and complex plant equipment procurement. As each procurement for such equipment has its own special features, appropriate modifications may have to be made to the listed items, and a thorough check made to ensure there are no omissions.

8.1   INVITATION TO BID

There are many terms used for inviting bids. They include Invitation to Bid, Request for Quotation, Notice to Bidders, and Notification to Contractors. The term *Invitation to Bid* is preferred as it is universally used and clearly describes the intent of this document.

Bid invitations for private procurements are usually mailed to bidders whom the purchaser has selected for their proven experience and reliability for the type of equipment to be procured. Attached to each invitation are the other three bidding documents. The invitation is either a standard preprinted fill-in form, or a one page letter inviting bidders to submit an offer to undertake the work to be covered under the contract.

Many government agencies funding public projects and those funded by international monetary lending authorities (such as, The World Bank and the Asian Development Bank) are required to publish invitation to bid notices in national and international newspapers and other selected publications. As already described in Chapter 3, the purpose of this invitation is to notify all parties who might be interested in submitting bids.

The invitation should be limited to information that will enable prospective bidders to determine whether they have an interest in bidding for the contract. As the instructions to bidders (see Section 8.2) accompanies the bid invitation, the invitation itself should be brief and restricted to:

- The name and address of the purchaser
- Name of the contract and contract number
- The name of the project and project number
- Location of the project
- Optional—a simple statement of the equipment being procured, which may read:

  > The design, manufacture, and delivery of ....
  > (*name of equipment*) and the services of advisors for installation and commissioning

- Request to advise the purchaser immediately whether the bidder intends to submit a bid
- If declining to bid, request bidding documents to be returned to the purchaser

8.2   INSTRUCTIONS TO BIDDERS

The *Instructions to Bidders* document sets out the instructions with which the bidders are required to comply for the preparation of their bid. When drafting the document, it is most important to fully understand the purpose of the document, that it is for preparation and submission purposes only. The instructions to bidders have no further relevance in the execution of the project after the contract has been signed. Therefore, any technical and/or commercial matter considered to be contractual should logically be included in the applicable contract document for bidding (see Chapter 3, Figure 3–1).

The overall framework of the information and instructions follows a similar pattern that can be applied to most equipment procurements. Details will obviously differ from one procurement to another depending on the purchaser's contract requirements as well as the nature, size, and value of the work.

It is accepted practice to require bidders to hold their bids valid for a fixed period (in number of days) after the closing date for the submission of bids. This period is normally referred to as the *validity period* and must be defined in the bidding documents. It usually ranges from 60 to 90 days, depending on the complexity of the procurement. Lengthy periods should be discouraged as bidders could submit higher prices than are appropriate due to uncertainties with respect to inflation and other economic factors. However, sufficient time should be allowed to permit the purchaser to perform a complete evaluation of bids, to negotiate with favored bidders, and to obtain approval from management, when required, to commit to the selected bid for procurement. If a bid selection has not been completed by the time the validity period has expired, the purchaser has to either:

- Negotiate an extension of time to the validity period with the favored bidders, *or*
- Reissue the bidding documents for rebidding

These matters are further discussed in Chapter 9.

As the validity period during which bids are kept open is limited, it is essential that clear instructions be given to bidders. Much of the purchaser's time is often wasted searching for bidders' responses to explicit requirements if the information has not been set out properly in the bidding documents. Providing clear directions to bidders also facilitates the bid evaluation process for those responsible for the evaluation.

By prescribing in the instructions to bidders the purchaser's requirements regarding the preparation and submission of bids, it is equally important to inform bidders of the policies the purchaser will adopt for evaluating bids. At the same time, both a purchaser drafting the documents and bidders submitting bids should be aware of local legal codes of practice concerning bid submittals and assessment of bids. Needless to say, the requirements for bidding should be strictly complied with by both the purchaser and the bidders. Bidders should only be obliged to respond to those requirements set out in the bidding documents. These requirements cannot be altered by the purchaser for any one bidder without all bidders being informed of the change.

The instructions to bidders should be contained in a separate document identified as such with headings and a clause numbering system. Typical series of instructions may be grouped under six headings:

1. General
2. Bidding Documents
3. Preparation of Bids

4. Submission of Bids

5. Opening of Bids and Bid Evaluation

6. Award of Contract

Most if not all of the following items need to be addressed in the instructions to bidders:

General

- *Name of the owner*— if the owner is not the purchaser
- *Name of the purchaser and address*
- *Contract number*
- *Type of contract*—lump-sum (fixed or subject to price adjustment), or cost-plus
- *Programmed date for award of contract*
- *Delivery to the named destination*—nominated location where equipment is to be delivered
- *Delivery program*—elements to be stated:
  - required time in weeks or date (ex works, carrier / alongside ship / on board ship or to site), *or*
  - bidder to state best possible delivery, *or*
  - purchaser's criteria for bid evaluation based on bidder's proposed delivery
- *Delivery responsibilities*—whether the purchaser or the contractor is responsible for providing any services, for example the use of a crane for unloading, storage facilities, and the like
- *Bidding period*—to be prescribed, but as a guide is usually at least 60 days for less complex equipment and approximately 90 days for major plant proposals
- *Cost for bidding*—that the bidder shall bear and be responsible for all costs relating to the preparation and submission of the bid
- *Subcontracting the work*—bidders who propose to subcontract the whole or major part of the procurement will not be considered
- *Contact name*—purchaser's representative and location for queries: name, title of person, address, telephone and facsimile numbers
- *Information and data to be submitted with bid*—making reference that it is contained in the separate document titled Content of the Bid
- *Language of bids*—that bids be written in the English language or in the language of the purchaser's country, unless specified otherwise. Submitted printed material may be in another language provided an exact English (or other as specified) translation is furnished with the material. All correspondence and other matters between the purchaser and the bidder shall be in the English language (or other as specified) unless specified otherwise in the contract documents.
- *Advertising and publicity*—that the purchaser reserves the sole right to all publicity and advertising in connection with the work

Bidding Documents

- *Charge for bidding documents*—bidders who have been invited to bid will be provided with a reasonable number of bidding documents free of charge. Additional copies of the documents may be purchased by contacting a nominated person at a stated location at a charge of $....................... per copy. A note should also state whether a non-refundable charge is made for all sets of documents supplied.

- *List of purchaser's bidding documents*—a listing of all the purchaser's bidding documents to be noted

- *Inspection of other related documents*—that suitable arrangement can be made with the purchaser giving a contact number

- *Clarification of documents*—contact name in the purchaser's organization if different from the above named person, to bring to the purchaser's attention any ambiguities, duplications or queries raised by the bidder and requiring clarification by the purchaser prior to bidding closure

Preparation of Bids

- *System of units in bids*—specify that the bidders offer is to be in the International System of Units having abbreviations SI for the Système International d'Unités as adopted by the General Conference on Weights and Measures or unless specified otherwise in the contract documents

- *Currency*—purchaser should state clearly in the bidding documents whether currency is to be quoted in the bidder's country where the bid is being prepared for submission or in the country of the purchaser's delivery site

- *Joint venture bid submission*—requirement that the bidder shall clearly state the names of the consortium companies bidding, the proposed arrangement for scope of work for each consortium company to perform, and the agreement for the management and execution of the work

- *Bid prices*—to be completed by the bidder in the schedules included in the Content of the Bid document

- *Contract price adjustment*—purchaser to state whether the contract price is to be fixed or subject to price escalation for labor and materials. The appropriate escalation formulae should be included in the contract documents

- *Tax*—whether the goods to be supplied are subject to or exempt from taxes, and who pays

- *Insurance*—who is responsible, purchaser or contractor

- *Bid validity period*—number of days to be stated from the date of bid closure. The length of the validity period will depend on the complexity of the procurement.

- *Partial bid*—a statement that such offers will not be considered if the bidder intentionally only bids for part of the work without first applying for the purchaser's approval to bid for part of the work. However, where say, all the bidder's completed schedules were left out unintentionally from one part of the bid, this becomes a nonconforming bid.

- *Nonconforming bid*—purchaser should state that a bid that has been improperly completed or does not comply with bidding documents may not be acceptable
- *Noncompliance features of the bid*—that these shall be clearly listed separately by the bidder in the bid. Acceptance of these is at the discretion of the purchaser.
- *Alternative bids*—a statement whether an alternative bid will be accepted.
    Note: Bidders are normally encouraged to submit alternative proposals if they believe the alternative will meet the specified performance requirements and at the same time offer a possibility of lower costs and/or other benefits to the purchaser. The statement on alternative bids must clearly indicate to bidders whether to submit either (1) a single bid that does not conform entirely to the specification but meets the purchaser's specified performance and objectives, *or* (2) the base bid that strictly conforms with specification, and an alternative that need not conform, but that would meet the objectives of the specification.
- *Bid form*—that the bidder has to complete and submit an "Original" and .... number of "Duplicate" copies of the form. Bidders are not permitted to alter the purchaser's wording in the bid form.
- *Bid security*—state whether a bid security is to be provided. If required, specify:
    - fixed amount or percentage of bid price
    - form of security
    - validity (days) beyond the validity of the bid
    Note: Bid security details are discussed in Chapter 13.
- *Endorsement of bids*—include format and signature of bidder's authorized person(s) on bid documents: the bidder shall sign both the Original and Duplicate bid (refer to "Submission of Bids" heading). If a corporation, the bidder shall affix its common or corporate seal or otherwise appropriately and formally sign the bid. The signatures of the bidder shall be witnessed. The name of each person signing shall also be typed below the signature.

Submission of Bids

- *Lodgement of bids*—Closing date, time and location
- *Telegraphic, telex, or facsimile bids*—will not be considered unless the bid is submitted by an overseas bidder and is received prior to bid closing time and date. Such bid shall note bid price and major technical and commercial features and any excep-

tions to the bid. The written bid must be received by the purchaser no later than seven calendar days after the bid closing date.

> Note: The purchaser must clearly define the above requirement in the bidding documents.

- *Bid submission requirements*—sealing and marking of bids: "Original" and .... number of "Duplicate" bids; whether separate for technical, commercial, priced, and unpriced. The sealed envelope of the bid is to be marked with the name of the bidder and the applicable contract number.

> Note: For smaller and less complex procurements, a duplicate may be sufficient and the purchaser's in-house reproduction can make as many copies as may be required. However, speedy and urgent review of a major procurement could involve a number of individuals analyzing different parts of a bid. Moreover, commercial parts containing pricing data may be treated as "confidential information," and have restricted circulation within the purchaser's organization.

Opening of Bids and Evaluation

- *Opening of bids and announcing bid prices*—advise as whether to be made in the presence of bidder's representatives and whether to be made public.

> Note: It may be mandated by law that bids are publicly opened and prices read out

- *Basis of bid evaluation and award of the contract*—advise that bids are evaluated on technical, financial, and contractual merits. They include guarantees of efficiencies of the plant or equipment and other characteristics of the design and operation of the plant or equipment, the monetary value of the work, and factors that affect the cost to the owner of the contract, such as maintenance requirements and operating costs.

- *Acceptance of bids*—it should be stated that the lowest or any bid will not necessarily be accepted. The purchaser reserves the right to accept any bid, to reject any bid, or to reject all bids.

- *Bid copy precedence*—that in the event of any discrepancies between the Original and the Duplicate, the bid marked Original shall prevail.

Award of Contract

- *Notification of acceptance of bid*—that the purchaser will issue a formal notification of acceptance to the successful bidder.

- *Formal contract agreement*—advise that the successful bidder will be required to enter into a formal contract agreement with the purchaser to execute the contract.

> Note: A copy of the contract agreement form should be included in the bidding documents.

8.3   CONTENT OF THE BID

The preparation of the bidding documents and the evaluation of bids that follow for major plant equipment procurements can become extremely costly and time consuming. Bidders should be fully informed of the necessary data concerning the contract and the precise information that is to be included in their bids. They should not be expected to furnish any more information than is called for in the bidding documents. Unfortunately, this approach has not always been followed in practice, and bidders have been compelled to request further details from the purchaser during the bidding period, or, the bidder makes a statement in the bid that "this is our understanding of your requirement(s)." In the absence of definite instructions, different bidders will submit different proposals, making the bid evaluation process exceedingly difficult. In these circumstances, deciding on the preferred bidder can lead to possible controversies.

In order to avoid wasting time searching for bidders' responses to explicit issues in bids, and to ensure that all bidders provide only the relevant information needed for bid evaluation purposes, the purchaser should compile a *Content of the Bid* document that lists the specific technical, contractual, and financial information the purchaser needs to know in order to complete a comparable and meaningful bid analysis. This information is normally arranged and presented in the form of questionnaire sheets, or *Schedules*, to be completed and submitted by each bidder with the bid. These schedules have the distinct advantage of forcing bidders to provide identical information.

Each schedule should preferably be designated as a consecutively numbered document that, on completion by the bidder, forms part of the bidder's offer, and after award of contract, of the contract documents. This information becomes an essential part of the bid, and a purchaser may demand authentication of each schedule requiring the bidder's signature or authorized signatory, the date, and the bidder's signature witnessed.

A sample range of schedules for technical, contractual, and financial aspects to be considered in making an appraisal is set out below.

8.3.1   Commercial

Lump-Sum Price Structure

Separate prices to be submitted for the following:

- For equipment offered within the purchaser's country
  - Price Ex Works[1]
  - Transportation charges
  - Insurance
  - Tax if applicable

[1]Incoterms are to be used as detailed in Appendix C.

- For equipment offered from outside the purchaser's country
  - Price FOB/FAS
  - Price CFR/CIF
  - Tax, if applicable
  - Import customs duties, noting customs tariff item, and customs duty rate
  - Foreign currencies exchange rates to be noted as at a specified date
- Essential spare parts
- On-site advisors, rates per calendar day or for a specified period for:
  - Installation
  - Commissioning
  - Performance testing
  - Personnel training
- Schedule of unit rates for extra work requested by the purchaser

Contract Price Variations

- Schedule of price variations for:
  - Local manufacture (labor and material)
  - Overseas manufacture (labor and material)
  - Transportation

Terms of Payment

- In terms of percentages of the contract price at milestone dates, *or*
- Progress payments

Delivery

- Delivery time in weeks, or by a date:
  - Ex Works, *or* FOB/FAS
  - On site
    Note: Time for delivery of imported items is to be clearly stated and be included in the work program
- Approximate shipping mass and the mass per unit
- Off loading at site, whether purchaser or contractor responsibility
- Site storage, if required, space and safe custody responsibility to be provided by purchaser or contractor.

Work Program

- Proposed work program:
  - Engineering, procurement, manufacture and delivery
  - "S" curve, if requested, showing time versus activities

Contractor's Capacity to Perform, if the Bidder Is not the Prime Contractor

- Contractor's company structure and organization chart
- Contractor's résumés of key personnel assigned to the contract, each detailing:
  - Present position and responsibilities
  - Previous positions held and responsibilities
  - Professional, trade, and other qualifications
- Contractor's past experience and completion of similar contracts, if the bidder is not the prime contractor:
  - Type of plant or equipment supplied and value
  - Name of project, owner, and location
  - Date of project completion
- Latest financial statement of the contractor including a copy of the company's latest annual report.

8.3.2   Technical

Bidder's Proposal and Drawings

- Technical description of proposal
- Data sheets and other questionnaire forms to be completed by bidders
- Drawing and technical data submittal requirements, schedule of scope, and dates to be confirmed

Warranty Performance of the Equipment

Nonconformity with the Engineering Specifications

- A statement by the bidder of conformance with the specification(s), *or*
- Listing of variations from specification(s), quoting clause number, reason, and comment
- List of items not covered by the equipment warranty (defects liability) due to being subject to normal operating wear

Alternative Proposal(s), if any, in Addition to the Conforming Bid

Details of the Contractor's and Subcontractor's Quality Control Procedures

Maintenance Tools and Devices

- List of maintenance tools and devices intended to be supplied by the contractor as part of the contract with the value of each item included in the contract price. The purchaser is not bound to purchase all or any of the items listed.

Subcontractors

- Schedule of subcontractors listing names and addresses of proposed subcontractors (noting for each, the exact work to be subcontracted).

Listing and Pricing of Essential Spare Parts

- List of essential spares based on .... hours of operation or other specified requirement, intended to be supplied by the contractor as part of the contract with the value of each item included in the contract price. The purchaser is not bound to accept all or any of the items listed.

List and Pricing of Optional Recommended Spare Parts

Note: For both the essential and optional recommended spare parts, the bid should contain the following information:
- Equipment identification for which the part is intended
- Part number and description
- Sectional drawings showing location, part number, and complete manufacturer's identification of each part
- Quantity of each part recommended for purchase
- Unit price, and total for the number of recommended parts including sales tax if applicable
- Mass—net and shipping
- Country of origin and delivery time
- Information as to interchangeability of parts with other equipment furnished on this contract, or previously furnished for the particular location

8.4   BID FORM

The *Bid Form*, sometimes also called the Proposal Form, is a document normally issued with bidding documents on which the bidder submits an offer to perform the works under the contract. When signed by the bidder, it is legally deemed an offer by the bidder to enter into a contract on the terms and conditions stated in the bid form.

The bid form is prepared in the form of a letter from the bidder and addressed to the purchaser. As offers worded by bidders are frequently incomplete and even meaningless, the bid form should be prepared by the purchaser with blank spaces for the bidder to fill in the required information, and sign the document. This will assure similarity in the preparation, presentation, and submittal of bids on an equal basis by bidders. The uniform arrangement of information will also facilitate the bid evaluation process. A sample bid form is shown in Section 15.1.4.

Bidders are not permitted to alter the purchasers' wording in the bid form.

Most of the engineering institutions and national standard documents committees, as well as many industrial organizations, have printed bid forms prepared with a standard text. The general format of the forms may vary depending on the type of contract, pricing information, and other legal requirements pertaining to the contract.

As the bid form has legal implications when completed and signed by the bidder, care needs to be taken when it is drawn up. It normally consist of the following key elements:

- Project identification by name of contract and the contract number to which the bid applies.
- Name and address of the purchaser receiving the bid
- Name of the organization and address of the bidder
- Acknowledgment and confirmation that all bidding documents were received and have been examined, including supplemental notices
- The bidder offers to undertake the work in accordance with the contract conditions for the sum of: $................. *(stated in words and numerals)* or such other sums as may be determined in accordance with the schedule of prices attached to the bid form
- Confirmation that in the event of acceptance of the bid, the bidder shall execute the contract agreement within .... *(number)* days or period from receipt of the notification of award of the contract or letter of acceptance
- Confirmation of the contract program (as per bidder's offer or to the purchaser's requirements), to be effective as from the date of the notification of award of the contract or letter of acceptance
- Confirmation of the validity period of the bidder's offer being .... *(number)* days as stated in provision number .... in the Instructions to Bidders from the date of bid closure, and that the bid may be accepted by the purchaser at any time prior to the expiration of that period
- Confirmation of the bidder's agreement that the purchaser is not bound to accept the lowest bid or any bid received by the purchaser
- Confirmation that if the bid is accepted, the contractor will provide, if required, a performance security in the sum of: $...................... *(stated in words and numerals)* for the due performance of the contract

- Bid form dated—day, month, and year
- Signature of the bidder "under seal" or "under hand" (see Equivalents and Meanings of Terms)
- Witness of bidder's signature

Where bids are publicly opened and bid prices announced, the bid form may also serve as an official acknowledgment with the following blank spaces added:

Certified bid opened at :   ....

date :   ....

time :   ....

In the presence of :   ....

....

9

THE BIDDING PERIOD

9.0   INTRODUCTION

The bidding period is the interval of time from the issuance of bidding documents to the specified bid closing date and time or, as sometimes defined, to the time bids are formally opened. The time allowed for bidders to prepare and submit their bids will depend upon the magnitude and complexity of the equipment being procured. Generally the period varies from 60 to 90 days. The bidding period includes any written extension of the date issued to bidders subsequent to the initial release of the bidding documents.

Once the bidding documents have been issued to prospective bidders, managing situations arising during the bidding period requires a very disciplined approach on the part of the purchaser. They include:

- Handling inquiries, discussions, and liaison with bidders prior to bid closing time
- Issuing bidding modification notices to bidders
- A bidder requesting to make a bid modification prior to the bid closing time
- A bidder requesting to withdraw the bid prior to bid closing time

Recommended procedures from the time of releasing the bidding documents through to receiving and opening the bids are discussed in the remainder of this chapter. It is recommended that the purchaser should have in place a set of in-house procedures, used by all purchaser's personnel for dealing with those issues discussed in this chapter, to forestall any criticism that could be leveled at the purchaser for unfair treatment.

9.1   ISSUING THE BIDDING DOCUMENTS

When the drafting of all the bidding documents has been completed, checked, and approved, a final review should be made to ensure that the documents form a coherent package for bidding purposes. Thereafter, each assembled package should be checked for the following:

- All documents are attached to the invitation to bid, and in particular, consistent with:
  - Date and time for bid closing including validity period of bid
  - Location for lodging bids
- All documents are correctly titled and identified, and have the appropriate revision number
- All document pages are in their correct location, and that none are missing

Setting the bid closing date on weekends or on days immediately preceding or following a recognized holiday should be avoided. Bid closing time should be selected between 2:00 and 4:00 PM local time, allowing for a morning delivery of bids.

The bidding documents packages are then sent to bidders noted on the purchaser's list of bidders, or in the case of open competitive bidding, those names obtained by advertisement.

9.2   ETHICAL PRACTICES AND LIAISON WITH BIDDERS

It should be clearly recognized that in competitive bidding for a procurement contract, all bidders must have equal opportunities and be treated equally for preparing and submitting their offers. Discussions and informal "chats" with individual bidders during the bidding period should be avoided. With a sound and complete set of bidding documents, there should normally be no need for any communications between the purchaser and bidders. Nor should conferences be necessary with bidders during the bidding period unless a major criterion has to be discussed that normally would not be specified in the bidding documents.

The following are recommended practices for the purchaser and bidders during the bidding period.

- Should the need arise, any liaison with a bidder should only be made through the purchaser's representative named in the instruction to bidders.
- Meetings and telephone discussions with bidders concerning any aspect of the contract should be avoided.
- Bidders should be requested to forward their queries in writing to the purchaser's representative for clarification.
- Under no circumstances should the purchaser negotiate a variation to any section of the bidding documents with a bidder.
- A formal notice concerning an amendment to the bidding documents must be sent promptly to *all* bidders as is discussed in Section 9.3.

9.3   SUPPLEMENTAL NOTICES—MODIFICATIONS TO BIDDING DOCUMENTS

If the purchaser has to modify a bidding document during the bidding period and prior to the official receipt of bids, the modified material must be transmitted to all bidders as quickly as possible.

This is accomplished by the purchaser issuing a *Supplemental Notice*. Modifications may be necessary for the purpose of correcting errors and omissions or clarifying duplications and ambiguities found by the purchaser or brought to the purchaser's notice by bidders during the bidding period. The supplemental notice is also a means by which added information is made a part of the bidding documents and may include changes to the scope of work and a change to the time and location for receipt of bids.

It should be noted that bidders do not take kindly to any requirement involving reworking of their already completed work. There have been known instances where a bidder has preferred to decline to bid rather than commit further company time and funds as a result of a purchaser's modification of a section in the bidding documents. Furthermore, issuing a supplementary notice close to bid closure time should be avoided. Such actions may necessitate the need to extend the bidding period, the implications of which are discussed in the following section.

An example of a supplemental notice form is shown in Figure 9–1 (page 9–4). Each form issued should be numbered in the order of issuance starting from 1. The document should include:

- Name and address of the purchaser as letterhead
- Project title and project number
- Contract number and contract title
- Supplemental notice number
- Date of issue
- Name and address of the bidder
- The name of the document that has been amended and the item that was altered
- Signature of the purchaser's authorized representative

Each supplemental notice is forwarded to all bidders via means of a *Document Transmittal Notice*. An example of a document transmittal notice form is shown in Figure 9–2 (page 9–5). The transmittal notice is presented in duplicate so that the duplicate copy can be signed by the bidder and returned to the purchaser to acknowledge receipt of the transmittal.

Where modifications have been made to any of the documents listed in the technical requirements form, or to the form itself (for example, the Issue Description in the revision

---

XYZ CORPORATION
2547 Tenth Street
ANYTOWN NJ 00001

SUPPLEMENTAL NOTICE No ....

Date of issue: ....

To: ....
              (*name of bidder*)

              ....
              (*address of bidder*)

              ....

On the Work for:

              ....
              (*name of project*)

              ....
              (*contract name and number, and name of equipment*)

the Bidding Documents are modified as follows:

Approved by: ....

FIGURE 9–1   Example of a supplemental notice form.

XYZ CORPORATION
2547 Tenth Street
ANYTOWN NJ 00001

DOCUMENT TRANSMITTAL NOTICE No ....

Date: ....   Contract No ....

To: ....

Attention: ....

Transmitted herewith are documents for:

( ) Approval   ( ) Per your request
( ) Preliminary   ( ) Information
( ) Reference only   ( ) Construction
( ) Revised as noted   ( ) Quotation/bidding
( ) Approved as noted   ( ) Other ....

mark "x" where applicable

Document No.	Revision No.	P-print T-transp. O-other	Number of Copies	Document Title

Please acknowledge receipt of this transmittal by returning one signed copy to the undersigned.

Issued by: ....

....
(*title*)

Received by: ....   Date: ....

FIGURE 9–2   Example of a document transmittal notice form.

block), the revised document(s) noted on the supplemental notice must be forwarded to each bidder.

The modified technical document package forwarded to each bidder consists of:

- Document transmittal notice (in duplicate)
- Supplemental notice number
- Copy of the revised technical requirements form
- Copy of the revised technical document(s), if it is an attachment to the technical requirements form

When lodging a bid, the bidder certifies that all amendments have been taken into account by completing the fill-in portions in the bid form noting therein the applicable supplemental notice number(s).

9.4   EXTENSION TO THE BIDDING PERIOD

If an impending modification to the bidding documents is required within five working days prior to the bid closing date, a decision must be made as to whether to inform all bidders by telephone and confirm by a facsimile transmittal, or to postpone the bid closing time. Due consideration must then be given to such postponement, with possible effects on:

- Delayed start-up and completion of the contract
- Subsequent additional costs that would flow from such a decision

9.5   BID WITHDRAWAL BY BIDDER PRIOR TO BID CLOSING TIME

It is accepted practice that a bidder may withdraw his bid and request the return of the submitted bid prior to the bid closing date or prior to the time of formal opening of bids, assuming that no bids have been opened previously. Such a request must always be submitted to the purchaser in writing. It is strongly recommended that the purchaser verifies that a request for bid withdrawal is genuine and issued by a person authorized by the bidder to undertake such action.

A bidder's request for a withdrawal of bid must be kept on record together with the purchaser's acknowledgment of receipt of the request.

Certain jurisdictions do not permit a bidder to resubmit a new bid for the same contract once the original submission has been withdrawn.

9.6 BID MODIFICATION BY BIDDER PRIOR TO BID CLOSING TIME

It is normally permissible to allow a bidder to modify his bid prior to the bid closing date or, if bids are to be formally opened, then prior to the date of bid opening.

There are some countries, however, whose legal codes will not permit a bid to be modified in any form or manner since once it has been submitted the bid must be held in force for the entire validity period.

Bid modifications must be in writing and submitted to the purchaser in the same manner as for the original bid submission, except that the envelope should be clearly marked: "Amendment to Bid." When the revised bid has been submitted, each page where a modification has been made to the original page of the bid or the complete document itself must be initialled or signed by a person authorized by the bidder. If a completely new bid is offered, the rules of signature (not initials) must be followed.

10

RECEIVING AND OPENING OF BIDS

10.0 INTRODUCTION

As discussed in Chapter 3, a bid submitted by a bidder is a legal binding offer in response to the purchaser's invitation to bid. If the procedures outlined in the previous chapters have been followed, the bid would include:

- The bid form, completed and signed by the bidder
- A bid security, if specified in the bidding documents
- A description of the equipment offered
- Content of the bid including completed commercial and technical schedules
- An alternative bid in addition to the conforming bid if the bidder considers this offer has advantages over the purchaser's specified requirements
- Any additional material that the bidder considers significant to further support the bid

The suggested procedure from the time bids are received by the purchaser through to the bid evaluation stage is discussed in this chapter. It is recognized, however, that it may be necessary to make certain modifications to suit the purchaser's processing procedure for the type of procurement in hand.

10.1 RECEIPT OF BIDS

It is the bidder's responsibility to ensure bids are received by the purchaser before the bid closing time. Bids should preferably be hand delivered or forwarded by a courier service. Where bids have been mailed, they should be clearly dated when posted and the official receipt retained by the bidder.

Bids received by telex or facsimile before the bid closing time should only be considered if they meet the instructions as have been stated in the instructions to bidders document. The condition of acceptance should be that an officially signed confirmation bid follows immediately. Should the confirmation bid differ from the telex or facsimile on important points, the bid should be rejected.

10.2   HANDLING LATE BIDS

The instructions to bidders should clearly state that late bids will not be accepted, even if a late submission is not the fault of the bidder. A late bid endorsed with the time and date received should then be returned to the bidder unopened.

The main justification for the rejection of late bids is the possibility that a late bidder might have access to prices made known after bids have been opened. In most cases, it becomes difficult for the purchaser to verify the excuse presented by the bidder for the late submission of the bid.

It is strongly recommended that the purchaser have clear in-house guidelines to cover procedures for handling late bids particularly if late bids are considered due to reasons of urgency or lack of acceptable formal bids.

10.3   OPENING OF BIDS

Chapter 8 referenced that the instructions to bidders should outline the procedure the purchaser will follow regarding opening of sealed envelopes containing the bids. Three methods commonly used when opening bids are:

1. Bids publicly opened and read aloud, including alternative offers, usually in the presence of a representative of each bidder. Statutory regulations for publicly funded projects and those financed by major banking authorities usually require strict adherence to this practice. A formal record is made of the bid opening time and date, location, as well as of each bid received, including the date of the bid and reference number.

2. Some of the larger companies form an in-house bid committee before bid closing time. Bids are opened in the presence of all members of the bid committee, all of whom sign and date each bid received. A preliminary bid comparison is made and signed by all members of the bid committee before the bids are passed on for a detailed evaluation.

3. Bids are opened after bid closing time by the purchaser's authorized person. A record is taken of the date and the bidder's reference number of each bid received. The purchaser keeps bid prices confidential and the bid contents will only be known by those assigned to perform the bid evaluation.

It should be noted that a bid price read out immediately after bid opening is not a final indicator of the merits of the offer for a major procurement. The quality of the bid can only be assessed after all technical, commercial, and financial features have been evaluated as discussed in Chapter 11.

Except when formally opening bids in public, the purchaser's disclosing details of a bidder's offer to another bidder would violate all known codes of ethics. The fewer persons involved with the bids before awarding a contract, the less chance there is of any accidental disclosure. A further word of warning is that all opened bids should be locked away after office hours as a security measure. Some bidder "representatives" will do their utmost to find out their competitor's prices. It is not uncommon to receive an unsolicited communication from a bidder after bid closing time stating "we have reviewed our proposal and are pleased to advise a price reduction by the amount of $....................." Needless to say, this type of communication should receive no further consideration.

10.4   MISTAKES, CORRECTIONS, AND REQUEST BY BIDDER FOR WITHDRAWAL OF BID

After bid opening, a bidder should not be permitted to alter a bid and resubmit it based on a claim of error or otherwise. It should be noted however, that court decisions in some countries have permitted error corrections in certain circumstances.

In the event of bids having already been opened and a bidder wishing to withdraw a bid after uncovering a genuine error or mistake in the offer that can be verified by the purchaser, withdrawal of the bid should be permissible. A bid withdrawal after bids have been opened is subject to the requirements of the applicable laws in certain countries and also to the legality of a return of the paid bid security in these circumstances. The bid evaluation process, as described in the next chapter, should proceed as though the withdrawn bid had not been received. The fact that the bidder did submit a bid and consequently withdrew it should be noted in the bid evaluation report.

10.5   REJECTION OF BIDS PRIOR TO BID EVALUATION

The instructions to bidders should state that the purchaser has the right to reject any or all bids received. Possible reasons for rejecting a bid before bid evaluation include:

- A bid not being prepared as stipulated in the bidding documents. The bidder may have failed to comply with the requirements detailed in the instructions to bidders.
- Bids received after bid closing time.

- A bid forwarded by telex or facsimile before bid closing time and the written confirmatory bid received later having different information, especially a lower price from that quoted in the telex or facsimile.

- Insufficient number of bids received for a competitive bid evaluation. Government agencies and many larger companies generally require at least three bids to be received before evaluation can take place.

- The purchaser can cancel a proposed project between bid closing time and award of contract, thereby rejecting all bids received. This type of situation highlights an important issue that bidders are normally not permitted to withdraw their bids once submitted whereas the purchaser is free to cancel the project before a contract is signed.

11

BID EVALUATION AND RECOMMENDATION REPORT

11.0 INTRODUCTION

The bids prepared and submitted in response to the invitation to bid must now be evaluated and a contractor selected to perform the work under the contract. Placing the contract solely on a low bottom line bid price in order to get the work started as quickly as possible without assessing the technical and commercial content of bids received will frequently in the long term cost more than a higher bid and cause delays in the completion of the work. An example would be the acceptance of a low-priced bid where equipment with grossly under-rated surface area for heat transfer has been offered but has not been checked. In most instances, this deficiency will greatly reduce the expected performance of the equipment as compared with higher-priced bids with a larger surface area. The bid evaluation therefore, is one of the most important activities in the procurement formation process and requires considerable efforts especially for large and complex equipment procurements. Yet many engineers appear not to be conversant with procedures for evaluating bids.

The evaluating effort embraces a technical appraisal of major features such as equipment design and operating and maintenance charges. The commercial considerations not only include the contract price but also matters such as delivery dates, terms of payment, and price variations due to cost escalation, all of which must be taken into account in assessing the purchaser's cash flow position.

Once the bid evaluation is complete, a recommendation of a contractor to be awarded the contract is usually presented in a report submitted to the purchaser's management for obtaining approval to commit the procurement.

It must be emphasized that the guidelines outlined in this chapter can vary in detail and scope depending on the size and complexity of the procurement.

11.1 EVALUATING BIDS

The process of evaluating bids consists of a systematic technical, contractual, and financial analysis of proposals submitted by bidders to determine which offer will be prove to be most economical when assessing the purchaser's overall objectives, taking into account equipment design and performance conditions, as well as costs and liability factors.

The following summarizes a systematic methodology that will establish a uniform assessment of all relevant factors that should be analyzed in determining the most advantageous bid, and avoids taking any bias to a particular offer.

- Undertake a preliminary check for arithmetical errors
- For overseas bids, convert bid prices to local (purchaser's country) currency at a common date
- Review bids for substantive technical responsiveness
- Select those responsive bids to be evaluated, minimum of three bids
- Proceed with a detailed bid analysis and evaluation to include:
  - Technical features
  - Commercial features
  - Financial appraisal
- Make financial adjustments to bid prices where required to bring all prices to a common basis for price comparison, such as, variable offers as against firm-priced offers
- Negotiate outstanding contractual issues with the favored bidders
- Select a recommended contractor to be awarded the contract

Recommended guidelines and procedures for evaluating bids are outlined in the following paragraphs.

11.1.1 Responsiveness

A responsive bid is one that conforms in all material respects with the terms set out in the bidding documents. A nonresponsive bid is one not conforming with the bidding documents to the extent that it is unacceptable and cannot be considered for award of contract. It should therefore be rejected. Conformance relates to both the submission of bids and the substance of the bid, and is aimed at eliminating bids that are not responsive to the requirements of the bidding documents.

The primary and essential purpose behind the insistence on responsive bids is that all bids are on an equal footing for bid assessment so that the integrity of the competitive bidding process may be maintained. Reasons for rejection of a bid for nonresponsiveness may include:

- Failure to submit a bid bond or deficiency in the bid bond (a bidder may be requested to make up a minor deficiency)
- Failure to bid for the specified equipment

- Failure to meet a delivery date specified in the bidding documents
- Failure to satisfy eligibility requirements
- Failure to submit a confirming bid within the specified bidding period

The name of any bidder who submitted a nonresponsive bid and whose offer was rejected, should be noted in the bid evaluation and recommendation report (see Section 11.5), stating the reason for the rejection of the bid.

Exceptions to commercial conditions should not necessarily lead to rejection of a bid. These exceptions could be subject to financial adjustments to make the bid conform with specified requirements.

Rejection of a bid is a serious matter. The reason for the rejection must be clearly explained on the bid analysis sheets and in such manner as to eliminate any cause for dispute.

11.1.2   Using Bid Analysis Sheets

It is essential that the bid analysis be more than merely making a copy of information taken from each bidder's offer. Possible errors in each bidder's offer should be checked, and construction, materials, and performance information should be verified against the engineering specifications and equipment data sheet requirements.

All responsive bids should be evaluated against the specification to establish conclusively which equipment offered by a bidder is most suited for the purchaser's intended application and service, and why it is so. It is equally important to establish why other bids, especially those that are less expensive, are not desirable or not accepted. To assist those making the appraisal, as much information as necessary should be tabulated in a logical way, leaving nothing of importance out. To do this, bidders' information should be analyzed using standard comparable evaluation forms. The format of these forms is so arranged that the principal technical and commercial (contractual and financial) features of each bidder's offer are listed in spreadsheets for quick and accurate comparison.

A suggested set of standard fill-in forms to be used is the *Technical Bid Analysis Sheet* shown in Figure 11–1 (page 11–4), and a *Bid Analysis Summary and Recommendation Sheet* shown in Figure 11–2 (page 11–5). Both forms are arranged in columns with one column allowed for each bidder. A secondary column within the bidder's column is for a "yes/no" to record whether the bidder has complied with the specified requirement.

11.1.3   Technical Features and Appraisal

The following is a sample list of technical features that would normally be assessed for a bid analysis and evaluation based on the bidder's completed technical schedules:

XYZ CORPORATION
TECHNICAL BID ANALYSIS SHEET

SHEET ___ of ___

EQUIPMENT

EQUIPMENT TAG No.

CONTRACT No.

DESCRIPTION	SPECIFIED	BIDDER Bid Ref: Date:		REMARKS	BIDDER Bid Ref: Date:		REMARKS	BIDDER Bid Ref: Date:		REMARKS	BIDDER Bid Ref: Date:		REMARKS	BIDDER Bid Ref: Date:		REMARKS
		M	S		M	S		M	S		M	S		M	S	
1																
2																
3																
4																
5																
6																
7	*This column lists*															
8	*the technical*															
9	*features to be*															
10	*evaluated*															
11																
12																
13																
14																
15																
16																
17																
18																
19																
20																
21																
22																
23																
24																
25																
26																
27																
28																

M S – Meets Specification Yes / No

DATE	BY	APPROVED

FIGURE 11–1 Example of a technical bid analysis sheet.

XYZ CORPORATION

BID ANALYSIS SUMMARY AND RECOMMENDATION SHEET

	SPECIFIED	BIDDER Bid Ref: Date:	BIDDER Bid Ref: Date:	BIDDER Bid Ref: Date:	BIDDER Bid Ref: Date:	BIDDER Bid Ref: Date:
EQUIPMENT						
EQUIPMENT TAG No.						
CONTRACT No.						
COSTS AND OTHER INFORMATION						
1 PRICE COMPONENTS – EX WORKS	Local Currency					
2 PRICE COMPONENTS – FAC, FOB, FAS	Local Currency					
3 EXCHANGE RATE (DATE)	To $ US					
4 EXCHANGE RATE (DATE)	To $ US					
5 TOTAL COST OF OVERSEAS COMPONENTS CONVERTED TO	$ US					
6 TRANSPORTATION – OCEAN	$ US					
7 TRANSPORTATION – LAND	$ US					
8 INSURANCE	$ US					
9 CUSTOMS DUTY	$ US					
10 CONTRACTOR'S SERVICES	$ US					
11 ESSENTIAL SPARE PARTS	$ US					
12 ADJUST FOR	$ US					
13 ADJUST FOR	$ US					
14 ADJUST FOR	$ US					
15						
16						
17						
18 TOTAL CONTRACT PRICE	$ US					
19 FIXED OR SUBJECT TO ESCALATION						
21 TERMS OF PAYMENT						
22						
23						
24						
25 COUNTRY OF ORIGIN						
26 DELIVERY TIME EX WORKS (WEEKS)						
27 DELIVERY TIME TO SITE (WEEKS)						
28 SHIPPING POINT						
RECOMMENDED BIDDER		REMARKS	REMARKS	REMARKS	REMARKS	REMARKS
RELEASED FOR PURCHASE	BUDGET ESTIMATE					
BY	CHECKED	APP'D	DATE			

FIGURE 11–2. Example of a bid analysis summary sheet.

11 – 5

Submitted Conforming Bid

- Equipment data sheet(s) completed:
  - Equipment manufacturer, type, make/model number
  - Performance data
  - Performance curves
  - Construction/materials
- Technical schedules completed
- Submitted calculations to be checked and to ensure correct factors have been used
- Appraisal of equipment for:
  - General suitability
  - Ease of installation
  - Ease of operation
  - Ease of maintenance
  - General environmental features
- Review of drawings submitted:
  - General arrangement, layout and overall dimensions, check space availability, and the like
  - Equipment sections
- Power demand:
  - Total installed power (kW)
  - Normal running load (kVA)
  - Peak running load (kVA)
  - Maximum starting load (amperes)
- Energy consumption:
  - Power consumption (kWh) per unit mass at full and partial production rates
  - Fuel consumption
- Other utility consumptions:
  - Steam
  - Water
  - Compressed air (instrument and plant)
- Operating cost adjustments for evaluation comparison (see Paragraph 11.1.6)
- Equipment warranties:
  - Performance (including items such as vibration and noise)
- Acceptance of listing of items not subject to equipment warranty—considered normal operational wear
- Material application check:
  - Weather protection adequacy
  - Surface preparation and painting
  - Heat and cold insulation

- Environmental considerations:
  - Noise
  - Pollution
  - Effluent treatment
- Electrical equipment appraisal
- Instrumentation and control equipment appraisal
- Scope of essential and operational spare parts:
  - Maximizing interchangeability of parts
- Scope of special tools and devices to be supplied
- Schedule of contractor's advisors for:
  - Installation
  - Commissioning
  - Training program for operator and maintenance personnel
- Assessment of contractor's capability to perform the work
- Extent of subcontracting
- Contractor's after-sales service and spare parts availability

Submitted List of Exceptions to the Specification(s)

Check listed nonconforming items and their effects on technical suitability and general acceptance

Submitted Alternative Bid(s)

Check technical operating and maintenance advantages and disadvantages

11.1.4   Contractual Features and Appraisal

The contractual appraisal is based on the bidder's completed schedules. Features listed will require checking for conformance with the bidding documents and include:

- Bid form duly completed, signed, and submitted with bid
- Bid bond furnished, if specified
- Confirmation of the bid validity period
- All documents submitted have been signed in accordance with the instructions to bidders
- Preliminary work program submitted
- Confirmation of time for delivery:
  - Ex Works or FCA/FOB/FAS in weeks
  - On site in weeks

- Point of shipment
- List of contractual nonconforming items, their acceptance or rejection
- Schedule of commercially acceptable subcontractors
- Listing of any licenses, design rights, and copyrights
- Insurance policy details presented; period and risks covered acceptable

11.1.5   Financial Appraisal

The financial appraisal will also be based on evaluating the schedules completed by the bidders. Features to be assessed include:

- Contract price, including:
  - Ex Works
  - Transportation and handling charges (off-loading and storage at site by the purchaser)
  - Insurance, if specified
  - Essential spare parts
  - Contractor's services at the site as specified
  - Unit schedule of prices for extra work requested by the purchaser
- Duties of customs information for overseas plant
- Currency exchange rate(s) for overseas procurement(s)
- Price structure:
  - Domestic procurements
  - Overseas components and customs duty rates
- Price fixed or subject to escalation with applicable formulae and indices for:
  - Domestic procurement
  - Overseas procurement
- Check terms and methods of payment, if asked for by bidders
- Check warranty terms
- Check insurance cover, terms of policy adequacy

11.1.6   Bid Adjustments

Once the bid evaluation is complete and all analysis sheets filled in, it may be necessary to make an adjustment to the cost summary of "apparent" low-priced proposals for cost extras needed to bring all bids to a common basis for price comparison to determine the lowest evaluated bid. Adjustments to bids may be required in the following areas:

Completeness in Scope of Supply

A minor omission in a bidder's scope of supply should not be cause for a rejection of the bid. Instead, a reasonable price should be estimated for the supply of the omitted item and be added where applicable to the bidder's price for comparison with other bids received for the particular item. The additional price adjustment should be based on the fair price of the omitted part and in line with the overall price. A review of the scope of work may be such that the design offered by the bidder could mean extra work to be performed by the purchaser. This work must be costed and added to the contract price.

Compliance with the Technical Specification

A similar occurrence to the one mentioned above, providing an item (considered as minor), but not as specified, should not be regarded as a major deficiency in compliance with the technical specification and therefore should not be cause for a bid rejection. An example is the case of the bidder having included access ladders instead of stairways to platforms. The cost of making good any deficiency should be added to the bid price concerned.

Adjustment for Efficiency, Losses, or Running Costs

For energy-consuming equipment, adjustment for the following is particularly important as these costs are real and recurring, and amount to substantial losses or savings throughout the life of the equipment.

- Efficiency for boilers, turbine-generators, pumps, fans, etc.
- Losses for electrical equipment (motors, transformers)
- Running costs for vehicles (mine haulage trucks), and other mobile equipment

The cost of energy should be calculated over the operating life of the equipment, discounted to present day value, and added to the bid price for bidder comparison.

The value to be used for calculating fuel costs and other prices should be the ones stated in the bidding documents for bid evaluation purposes.

Calculation of efficiency and fuel losses or gains should be for a fixed period of time of expected equipment operation (for example, 5 years) based on warranted performance figures submitted by bidders.

Examples include:

Each kilowatt of electricity required carries a cost of $ ....

Each additional kilowatt of power is worth $ ....

Each one percent below the warranted performance carries a penalty of $ ....
Each one percent above the warranted performance is worth $ ....
A reduction of .... kWh per tonne of production is worth $ ....

Terms of Payment

Terms of payment are generally finalized in the negotiation stage with the three favored bidders. However, adjustments should be carried out and an amount added to the bid price to compensate the bidder requesting an initial larger payment than the purchaser was prepared to accept, or for a payment that was sought earlier than scheduled. A suggested method for calculating the sum for compensation purposes covering these variations is to use commercial interest rate values that are applicable or relevant to the circumstances in question.

11.1.7   Contractor's Resources and Facilities

Of particular importance is the availability of the favored bidder's facilities and assigned key personnel for meeting the contract program. A frequent cause for not achieving the specified completion dates is the lack of available personnel or facilities due to the contractor's commitments on other projects. The purchaser must therefore review the workloads of the favored bidders and their ability to handle the procurement under consideration.

From the bidder-completed schedules (see Chapter 8), the purchaser should check:

Standard Company "Boilerplate"

- Overall services offered by the company
- Company organization structure
- Specific areas of technology
- Company's facilities for improving current components and developing new products
- Company's financial position, latest audited financial statement

Talent Inventory

- Evidence of good organization and management practices
- Qualification of personnel:
  - Permanent staff or the need to utilize contract staffing
  - Credentials of project manager and key personnel

- Engineering and computerized design capabilities
- Support groups:
  - Cost control
  - Planning and scheduling
  - Inspection and expediting
- Construction personnel:
  - Liaison with client
  - Language skills for overseas contracts

Adequacy of Facilities

- Fabrication facilities:
  - Shop management and supervision
  - Standard of plant and machinery
  - ISO 9000 registered for quality assurance

Experience in Similar or Related Fields

- Type and size of projects previously undertaken
- Experience with practices and procedures to perform the work
- Geographical or climatic experience for overseas contracts
- Technical advisors available to meet unforseen problems

Record of Past Performance

- Commitments generally held
- Reliable performance of installation
- Cooperation with clients
- Cost overruns attributed to underestimating, omissions, unanticipated problems
- Company capability to solve its own problems without seeking outside assistance
- Liaison with a purchaser's representative for on-site activities

11.1.8   Risk and Liability Analysis

Both the purchaser and the bidder (contractor) must accept the fact that there always can be an element of risk with any major equipment procurement. Industries in general are becoming increasingly capital intensive, and substantial investments are committed to improve existing products as well as to create new production capacity. It is here where both the seller's and buyer's management are required to quantify the risks involved. In the procurement process, risk abatement comes down to the thoroughness of practical planning and strict adherence to specific requirements. The purchaser should undertake a thorough review for possible minimizing of risk, as any major risk associated with the procurement should be a sufficient deterrent to bid acceptance.

Typical risk factors the purchaser should consider include:

- Will the contractor have difficulty with the procurement in areas such as technical design or penalties on performance or delivery?
- Can the contractor deliver the equipment within the specified contract delivery date?
- Are there sufficient safeguards built into the contract to ensure subcontractors and major material suppliers will comply with their supply and delivery commitments to meet the contractor's work program?
- Based on a very low contract price, can the contractor's engineering and manufacturing sections deliver the equipment without encountering financial difficulties?
- Has the bidder purposely bid a price that is lower than its projected costs because of knowledge or expectation of procurement of additional units in the future? Will the bidder be willing to sustain the loss on the first contract in the return for engineering and production experience that would place the company in a more favorable place for future procurements?
- Is the company willing to bid a low price to enter into a contract that will result in a loss in order to keep key people in the workforce employed during a slack period?
- Does the contractor have a past record of cost overruns?

11.1.9   Points Weighting System for Bid Comparison

When all the bids have been evaluated and cost adjustments made to bring respective bids to a cost comparable basis, it might still not be possible to make a final decision for selecting the successful bid. A more accurate assessment involves applying a point system to weight a series of selected features of the bids.

A weight can be assigned to each of the following performance factors:

- Understanding of requirements
- Excellence of technical design
- Completion of the work
- Management and assigned key personnel
- Quality and performance reliability assessment (taking into account maintenance costs)

Each bid is then graded based on the bidder's present and past record of strengths and weaknesses to determine how much of that weight can be added to obtain a final decision as to which bidder is to be awarded the contract.

The evaluation factors may include but not be limited to:

- Management's and key personnel's strengths and weakness
- Product and application knowledge
- Accuracy and completeness of bid submitted
- Prompt start of work after award of contract
- Quality of fabrication, inspection and quality control
- Meeting delivery date
- Product operational performance
- Incurred maintenance costs
- Drawing and technical data submittals:
  - Meeting submittal dates
  - Quality of information submitted
- Quality of installation and operating and maintenance manuals
- Technical quality of contractor's on-site advisors
- Sales service response
- Past record of satisfactory work performed

A typical example is shown in Figure 11–3 where an established maximum of 1,500 points is distributed among the above mentioned five performance factors in accordance with a premeditated ranking of importance. For this example, four bids were considered. Bidder A's offer was then graded on a scale of 1 to 10 under each factor. The grade score

Performance Factor	Assigned No of Points	BIDDER A's BID SCORING		
		Grade Score	Grade Percent	Grade Points
Understanding of requirements	500	10	100%	500
Excellence of technical design	400	6	60%	240
Time for completion of the work	300	8	80%	240
Management and assigned key personnel	100	5	50%	50
Quality, performance, and reliability	200	7	70%	140
Total score	1500			1170

FIGURE 11–3   Typical bid grading and scoring.

was then computed into a grade percentage to obtain a bid grade point score for each performance factor. The summation of the grade points received becomes a summary of the strengths and weaknesses of the bid.

A summary comparison of grading and scoring of the four hypothetical bidders is shown in Figure 11–4.

Performance Factor	Bidder A	Bidder B	Bidder C	Bidder D
Understanding of requirements	500	450	350	300
Excellence of technical design	240	400	360	250
Time for completion of the work	240	150	300	150
Management and assigned key personnel	50	70	50	60
Quality, performance, and reliability	140	160	120	180
Total score	1170	1230	1180	940
Average score [of 1500 points]	78%	82%	79%	63%

FIGURE 11–4   Example of comparison of performance scoring between four bidders.

11.1.10 Alternative Bid Analysis

For certain major equipment procurements, the purchaser may consider that there are other specifications that could meet the performance requirements and at the same time offer the possibility of lower costs and other benefits. The purchaser must also recognize that reputable equipment suppliers have their own expertise, skills, and experience. Furthermore, the purchaser may not always be aware of a supplier's latest development of new and advanced technology for a particular product. He should then clearly indicate in the instructions to bidders document the possibility of submitting alternative bids either as:

1. A single bid that does not conform entirely to the specification but meets the purchaser's performance requirements and objectives, *or*
2. To submit two bids, one of which conforms strictly with the specification, and the other that need not conform, but which would meet the objectives of the specification

Features of alternative bids may be discussed with bidders during the negotiation stage as described in Section 11.2.

11.2   NEGOTIATING WITH SELECTED BIDDERS

Following the completion of the bid evaluation, it is accepted practice for the purchaser to conduct negotiations with bidders who are on the "short list" (usually no more than three bidders) before making a final decision for awarding the contract. Typical items for negotiation may include:

- Review of the overall responsiveness of the technical part of the bid
- Review and appraisal of any of the bidder's listed exceptions to the specification(s)
- Review of an alternative bid offered by the bidder and the merits of that bid compared with the purchaser's specified requirements and performance warranty
- Review of the contract price breakdown
- Terms of payment
- Transportation details
- Contract price variation provisions
- Time for delivery, and possible accelerated delivery for early contract completion with a possible increase in the contract price
- Acceptance of liquidated damages
- Review of the drawing and technical data requirements form concerning scope of documents to be submitted and submittal times or dates
- Warranties
- Finalize special tools and spare parts requirements
- Extent of subcontracting sections of the work

Items for discussions must not center around a change in the scope of work. This should only be done when all bidders are notified of such change.

An important point to remember is that if the bid evaluation process has taken longer than expected, time may be limited before the deadline date of the bid validity. These meetings should therefore be well organized in advance, or otherwise it may be necessary to negotiate an extension to the validity period. Thorough planning and preparation by the purchaser's team attending each meeting are essential, including decisions for:

- Who will prepare the agenda for the meeting
- Who will preside over the meeting (usually the project manager)
- Who is authorized to speak and when
- Who will make the decisions
- Who records the minutes of the meeting

Bidders who are requested to attend should be fully informed beforehand of the agenda to be discussed to assure the availability for attendance of individuals with the necessary resources to carry out the negotiations. The purchaser should be aware that these bidders will assume they are on the short list. Having reached this stage, it could give them an opportunity to offer an improved bid price and/or any other aspect of their offer in order to secure the contract. Such offers would normally be subject to rejection but in the real world, acceptance occurs.

A discussion as outlined above need not be a round-the-table meeting but can also be conducted by telephone with a conference call connection. In all cases, the items discussed must be carefully minuted and placed on record. A document placing on record the salient items discussed is a *Written Confirmation* form, which is shown in Figure 11–5 (page 11–17). The form notes:

- Contract number
- Whether the minutes record a meeting or telephone conversation
- Date of meeting or telephone conversation
- Location where the meeting was held
- Purpose of the meeting
- Name of the person who recorded the minutes
- Names of people who attended the meeting or who were connected by telephone
- Items discussed and results agreed upon and any other actions required

A copy of the written confirmation signed by a purchaser authorized person (usually the project manager) should be forwarded to the bidder as soon as possible after the meeting with a request to confirm in writing that minutes presented are a true record of the items discussed and decisions agreed upon. Copies of the minutes are then distributed within the purchaser's organization.

11.3   CHECKLISTS FOR BID EVALUATION

To avoid unnecessary duplication, a series of bid evaluation checklists for a sample equipment procurement are contained in a Part 5, Section 15.4.2.

11.4   PURCHASER'S RIGHT TO REJECT ALL BIDS

The purchaser may exercise the right to reject all bids. This occurs only in very extreme situations where, for example, bid prices are far beyond the purchaser's financial capability to fund the procurement (where the purchaser is at fault for mismanaging the budget

XYZ CORPORATION

WRITTEN CONFIRMATION

CONFIRMATION No   **CONTRACT No**   **SHEET** 1   **of**

CONFERENCE / TELEPHONE   **DATE**

MEETING HELD AT :

SUBJECT :

RECORDED BY :

PRESENT :

ACTION BY	

DISTRIBUTION :

FIGURE 11–5   Example of a written confirmation form.

estimate) or the total project has either been canceled, re-engineered, or rescheduled beyond the bid validity period, or not enough bids of an acceptable nature are received.

The purchaser must fully appreciate and understand that for the bidder to engineer, estimate, and prepare a bid is an extremely costly exercise that is borne by an overhead account. Because of this, not all bidders may be prepared to rebid at a later stage and extreme care should therefore be taken to avoid any requirement for rebidding a project.

11.5   PREPARATION OF THE BID EVALUATION AND RECOMMENDATION REPORT

As soon as the bid evaluation is complete and the bid comparison sheets have been checked for the selection of the bidder for award of contract, it is common practice to prepare a bid recommendation report to obtain approval before committing a major equipment procurement.

The recommendation report may have to be directed to:

- A person in the purchaser's company or government agency, at a level of authority (from manager to board chairman) in accordance with the contract price of the proposed commitment
- The plant owner's management if the purchaser has been assigned to develop the procurement on behalf of another party
- A principal of the organization financing the contract

The report should be produced in clear and simple language containing all necessary information and summarizing the principal reasons why the successful bidder was chosen. It should be noted that not everyone who reviews and approves the recommendation has technical qualifications, and many may be more concerned with the financial issues of the procurement. The intent of the report, therefore, is to convince all members of the approving authority that the evaluation has been performed in a responsible manner and that a correct recommendation has been made for the selection of the successful bidder.

The evaluation and recommendation report usually includes:

Covering letter or interoffice memorandum with attached bid evaluation summary sheet(s), and the following information:

- The name of the contract and the contract number
- A short description of the extent of the procurement
- Bidders who were invited to bid
- Bidders who declined to bid

- Bidders whose bids were rejected and reasons why
- Comparison summary of (say three) preferred bidders
- Name of contractor recommended for award of contract
- Difference between the bid price and the estimate
- Validity date of offer

The basis on which the recommendation was made were:

- Most favorable price
- Most favorable payment terms and conditions, or compliance to terms and conditions
- Acceptable time for delivery
- Shown production capability to comply with the contract program
- Specifically accounted for all items and services to be supplied under the contract
- Has shown for company organization, personnel, and facilities:
  - Evidence of good organization and management practices
  - Key personnel with appropriate qualifications and experience
  - The working organization for the contract with the key people in place
  - Manpower projections, program loading, and resources capabilities
  - Adequacy of all necessary facilities
  - Proven experience in similar or related fields to the proposed work under the contract
  - A sound record of past performance
- Conformance with the technical specification
- Technical considerations included:
  - Completeness of the equipment offered
  - Demonstrated major design features and where economics are involved:
    Fuel and power costs including energy efficiency
    Load factors
    Environmental concerns
    Ease of installation
    Projected operating and maintenance costs
    Key components are interchangeable, minimizing stores inventory
    Ready availability of spare parts, and after sales service
- Understands the purchaser's approval procedures
- Understands relationship requirements with subcontractors

12

FINALIZING CONTRACT DOCUMENTS AND AWARDING THE CONTRACT

12.0 INTRODUCTION

Chapter 11 discussed the bid evaluation process leading to the selection of a successful contractor to perform the work under the contract. It was assumed that managerial authority had to be given, and was received, before the procurement could go ahead.

This chapter outlines procedures for:

- Issuing advanced contract notifications to the successful bidder
- Revising and updating the contract documents to evidence the contract
- Awarding the contract
- Notifying unsuccessful bidders
- Signing the formal contract agreement

A frequent practice to be firmly discouraged is for someone in the purchaser's organization, acting without authority, to telephone the successful bidder to say that his company is the preferred bidder and that he will be getting the contract. It must be understood that no binding contract is formed until the successful bidder has been notified in writing by the purchaser of an award of contract, and that the bidder has acknowledged the notification. The purchaser must be fully aware of the dangers of allowing the contractor to proceed with work under the contract ahead of any formal contract documentation. Such instruction can place him at a considerable disadvantage, and weaken a negotiation position should any item still be subject to agreement between the parties.

In passing, it is not uncommon for the purchaser to update the contract documents *before* submitting the bid evaluation and recommendation report to management for approval. This procedure runs the inherent risk that management may disagree with the recommendation, and directs that another bidder be awarded the contract, and with this comes a resultant loss of time. This seldom happens, nor should happen, but there may be situations where management is compelled to authorize awarding the contract to another bidder.

Strong pressure is applied by management at times to get the work started immediately before the contract documents have been finalized and the contract award notification forwarded to the successful bidder. It is suggested that the purchaser's problem could be solved by issuing the contractor with an advanced notification either by a letter of intent, or a notice to proceed thereby avoiding any delays to get the work under way. It is essential that both the purchaser and the contractor know and understand precisely what each of these documents means, and they are further discussed in this chapter.

12.1   ISSUING ADVANCED CONTRACT NOTIFICATIONS

Where there is an immediate need for work to start before finalizing the contract, it is usual to give the successful bidder advanced notification in writing that the purchaser intends to award the contract to that bidder, or, with no contractual commitment, authorizing him to incur specific preliminary expenses. Extreme care must be exercised to first ascertain under the country's contract law whether such notification does in fact constitute a legal binding contract.

12.1.1   Letter of Intent

A *Letter of Intent* is an advance notice to the successful bidder to advise that the purchaser "intends placing a contract with your company for . . ." The letter is a simple means of committing the purchaser to the agreement without negotiating the final terms and conditions of the contract. The use of the word "intent" would make it seem clear that the letter should not be a legal binding agreement since it excludes the essential information that would normally be set out when notifying the bidder of a formal acceptance of his offer. Hence, extreme care must be exercised with the legal interpretation and implication of the "intent." Laws in some countries have mandated that a letter of intent becomes a legal binding undertaking and that the purchaser must subsequently establish the contract as intended.

Even where the letter of intent is not legally binding, it will be interpreted by the bidder that the purchaser has a moral obligation to award his company the contract. The purchaser must therefore exercise extreme care that he does not slip on any negotiation leverage as outstanding matters move closer to an agreement.

The letter should be as detailed as possible, and must make it absolutely clear as to what is intended to be accepted, whether it is work in the conforming bid, the conforming bid with amendments, or an alternative bid submitted by the bidder. It must also be made clear that the letter of intent is a nonbinding document, and is not an authorization or instruction to proceed with any work. Its purpose is simply to inform the contractor to start planning the work and organizing the taskforce to be assigned to the contract. The contractor must accept the risk that any expenditure incurred for such preliminary work based on the letter of intent may not be recouped especially if the purchaser decides not to proceed with the contract (if permitted by law).

12.1.2   Notice to Proceed

In some cases, the purchaser may wish to proceed immediately with the work while outstanding items (a limited few) are still subject to negotiation by issuing a *Notice to Proceed* to the successful contractor. The notification is in fact a limited version of the notice of award (see Section 12.5).

    The notification may be worded in such a manner that with the contractor to start work on the contract the purchaser also wishes to inform the contractor that a notice of award of contract will be issued as soon as the negotiations are complete. The notice to proceed must clearly specify what part of the contract the contractor is to proceed to work on, and the limit, if any, of expenditure to be incurred. It is also important to specify the terms and conditions for work to be performed under this notification. The contractor should preferably be reimbursed on a negotiated cost-plus contract basis for labor and material. After the actual award of contract, amendments will have to be made to the contract price for payments made to the contractor for work that has been completed.

    The following is a typical form before the issuance of a formal procurement commitment.

Example of a Letter from Purchaser to Bidder

XYZ CORPORATION
2547 Tenth Street
ANYTOWN NJ 00001

Ref. OC/HD/N1712

(*date*) ....

Furgasonn Construction Company
1793 Main Street
Smithfield CO 00023

Attention: Mr. Joshua T. Jones, Senior Vice President

Gentlemen:

Devin River Mine Project
Contract No N-3721, Transportable Conveyors

With reference to your Proposal 589/T dated ...., we are pleased to inform you that your company will be awarded the contract for this work subject to agreement being reached between us on the matter of ....

As we wish to commence work on this contract immediately, we authorize you to begin design work in accordance with such instructions as you will receive from Mr. ...., our Project Engineer, up to a value not exceeding $ .... priced at the schedule of hourly unit rates set out in your proposal.

We reserve the right to terminate your design work at any time, and will reimburse your company for work carried out up to the date of termination.

Kindly acknowledge by facsimile, receipt and acceptance of this Instruction to Proceed, and thereupon to put the work in hand without delay.

Yours very truly,
XYZ CORPORATION

T.W. Maybolt, Jr.
Project Manager

The date shown on the notice to proceed must be qualified by a statement noting whether this becomes Day 1 when the contract comes into force, or how the notice will affect the contract program. This issue is of relevance should liquidated damages provisions be included in the contract for late delivery of the equipment.

The notice to proceed should be subject to the contractor's written acceptance. It will always be necessary to issue a notice of award and this should state that it replaces the notice to proceed and should clearly specify those matters that concern the remaining portions of the contract.

12.2   UPDATING THE TECHNICAL SPECIFICATION

Documents in the technical specification that must be brought from an "issued for bids" status to either "issued for purchase," or "revised as noted and issued for purchase" status are:

- Engineering specification(s)
- Equipment data sheet(s)
- Reference drawings, if revised, or no longer required
- Miscellaneous technical information relevant to the procurement, if revised
- The technical requirements form

Each equipment data sheet that was issued for the bidder to complete will have the successful contractor's information added in the fill-in spaces. The data sheet is then revised to "issued for purchase" and assigned the next revision number, usually from Revision 0 to Revision 1.

The drawing and technical data submittal requirements form originally issued for bids with Revision 0 status will remain Revision 0 unless a change has been made to the form, that is, a document submittal time has been amended. In the event of the latter, the form is marked up where required and updated to Revision 1.

Care must be taken to ensure all other technical documents noted in the Attachments block of the technical requirements form, such as reference drawings, reports, and chemical and/or property analyses, are no longer left in an "issued for bids" status. Normally, most of these documents will remain unaltered. However, information and data that is no longer relevant to the procurement should be deleted from the technical requirements form.

Revising the Engineering Specification for Purchase

An engineering specification may not require any alteration other than revising the issue description and revision number status. Any change to the specified scope and/or requirement should have been discussed and agreed on by the parties during the precontractual negotiations and be recorded in a written confirmation. Such changes may include:

- Altering purchaser or contractor responsibility for the provision of an item
- Altering a design requirement for the equipment
- Where the name of a supplier for an equipment component has been specified and "or equal" noted, the "or equal" *must* be replaced with what will be supplied under the contract

Typical revisions to an engineering specification from bidding to purchase status are shown in Figures 12–1 (page 12–12) and 12–2 (page 12–13).

Revising Equipment Data Sheets

Revising each equipment data sheet involves copying the fill-in information from the successful bidder's completed data sheet onto the purchaser's project data sheet, and revising the document from the "issued for bids" status to "issued for purchaser." The sheet should then be carefully checked, as it is an important part of the technical specification. The correctness of the document is the purchaser's responsibility.

A revised equipment data sheet with the successful bidder's information added is shown in Figure 12–3 (page 12–14).

Checking Reference Drawings

It is important to remove from the technical specification any reference drawing that was issued solely for bidding purposes and has no relevance in the contract. Other reference drawings listed in the Attachments block of the technical requirements form will remain unaltered, unless revised as required, to form part of the technical specification to be issued for purchase.

Miscellaneous Technical Information Relevant to the Procurement

This section is usually a series of exhibits to the technical specification and includes such information as records of meteorology data prevailing at the project site, geological and geotechnical logs from drill holes, water quality analysis, raw coal properties, and soil analysis data.

Normally, this information is not altered from the bidding to the purchase stage since it contains essential data required for the design of the equipment.

Updating the Technical Requirements Form for Purchase

An example of the updated technical requirements form from the "issued for bids" to "issued for purchase" status is shown in Figure 12–4 (page 12–15).

12.3   FINALIZING THE COMMERCIAL DOCUMENTS

Finalizing the commercial conditions usually includes the following:

- Amendments to the general (standard) conditions
  It is most unlikely that there will be changes to these conditions as set out in the bidding documents.
- Completing the attachment sheets to supplementary conditions for:
  - Name and address of contractor
  - Contract prices
  - Time for delivery
  - Terms of payment
  - Method of payment
  - Arrangement for payment in foreign currencies
  - Rates of exchange for the purpose of the contract
  - Contract price adjustments—formulae and method of calculating changes in costs
  - Maximum liability of the contractor to the purchaser
  - Changes, if any, to liquidated damages provisions
- Attaching the successful bidder's completed commercial schedules to the supplementary conditions
- Attaching the successful bidder's bid form to the supplementary conditions

12.4   FINALIZING THE SELECTED BID

The purchaser must never alter bidder's documents. After bid evaluation, the purchaser is responsible for checking that the successful bidder has incorporated revisions and resubmitted his bid, if changes have been made, together with the completed technical schedules (which now become part of the submitted bid), in accordance with the items that were agreed between the parties during the precontractual negotiations, and that were recorded in the written confirmations.

Revisions in the bid documents should be dealt with in the same manner as the purchaser's documents, that is, a triangle with the revision number and the corrected information added, rather than have the original document totally rewritten to incorporate the change(s). This will facilitate the purchaser's efforts for checking the bid in its "as bought" status. It becomes the purchaser's responsibility to ensure the correctness of the bidder's documents.

12.5   ISSUING THE NOTICE OF AWARD—LETTER OF ACCEPTANCE

The normal procedure for notifying the selected bidder of the purchaser's acceptance of the bid is a *Notice of Award* (term used in the United States) or a *Letter of Acceptance* (term used in other countries) signed by the person authorized by the purchaser. The purchaser's acceptance of the bid will be on the basis of the original bid or an original bid revised or amended to incorporate items that were negotiated and agreed by both parties.

The notice of award is an acceptance of the contractor's offer to enter into a contract with the purchaser. Normally included with the notice of award are copies of the updated contract documents. The notification gives the contractor a certain period of time within which to review the proposed contract and confirm, or state otherwise, acceptance of the document. To avoid doubts as to the contractor's actual acceptance of the proposed contract, the notice of award is usually forwarded to the contractor in duplicate with the duplicate signed by the contractor as a formal acknowledgment of the proposed contract and returned to the purchaser. The contract becomes legally enforceable by both parties at the date the notification is issued to the contractor or by the date noted in the notice of award.

Where the contractor disputes the proposed contract, the contractor must advise the purchaser in writing of the reasons, and in the timeframe allowed. This may require correction and negotiation to clarify and usually the notice of award would be withdrawn and a fresh notice issued with revised dates when agreement is reached.

The contractor may only dispute matters already under negotiation and no new items may be introduced as a reason to dispute the proposed contract.

It is important that the notice of award includes the basis on which the bid has been accepted by the purchaser, and be signed by an authorized representative of the purchaser.

The notice might include:

- Contract price
- Bid reference number and date
- The broad details of the equipment to be purchased
- The name of the purchaser's project manager to whom all communication shall be directed
- Confirmation that the date of notice of award is the date of acceptance of the bid, that is, Day 1 of the contract or as noted otherwise
- A statement whether work is to commence immediately or as noted otherwise
- Number of days for review of proposed contract

A further statement should be made that the bid has been accepted subject to the following conditions:

1. That within .... days of the date hereof, the contractor shall execute a formal contract agreement.
2. If required, that within .... days of the date hereof, the contractor shall forward copies of the executed performance bond (and other bonds as specified), and certificates of insurance that are necessary prior to the actual commencement of the work.

Should a performance security be required, the following paragraph shows a suggested way to phrase this requirement:

> The Security pursuant to Article .... of the Supplementary Conditions, the sum of $US .... shall be furnished to us within thirty (30) days within the date hereof in a form as noted in clause ....

Until the formal contract agreement is signed by the parties, the notice of award or letter of acceptance establishes the formal contract.

A sample notice of award of contract is given in Part 5, Section 15.5.1.

12.6   NOTIFYING UNSUCCESSFUL BIDDERS

Bidders should not be told the status of their offer until a firm and binding contract has been established. There may be last-minute problems or misunderstandings resulting in the selected bidder declining to accept the contract and withdrawing the bid. As soon as the contractor's acceptance has been received, unsuccessful bidders should be immediately informed by letter that their bid has not been successful and their bid security returned (if this was a bidding requirement).

There will be the usual approaches by unsuccessful bidders seeking information on "Where did we go wrong?" It remains the purchaser's discretion whether or not to furnish explanations.

A sample letter of bid rejection is given in Part 5, Section 15.5.2.

12.7   SIGNING THE CONTRACT AGREEMENT

Although in many countries the notice of award or the letter of acceptance will cause a binding contract to be formed for the execution of the work, it is common practice for the purchaser and/or the contractor to confirm the contract by requiring a formal agreement, hereafter referred to as the *Contract Agreement*, to be entered into. It should be made clear in the instructions to bidders document whether or not a contract agreement is required by the purchaser (see Chapter 8). The formal signing of the contract agreement by both the purchaser and the contractor emphasizes the importance and formality of the contract and is the final stage of the contract formation process. Furthermore, the contract agreement is a written confirmation that all items that were subject to negotiation have been consolidated into the contract documents exactly as finally agreed to.

Like other contract documents, the contract agreement is prepared as an integrated set of contract documents. It should be noted therefore, that while many people refer to the "Agreement" as the "Contract," it is technically incorrect to do so since the contract includes the contract agreement and all the other contract documents.

Assuming a contract agreement is required, the sample form should be included with the bidding documents. Sample forms of a contract agreement are included in many standard contract conditions. Alternatively, the document is drawn up by the purchaser's legal advisers to incorporate respective promises of the parties and specifying other considerations. Keeping the document as simple as possible is recommended leaving negotiated technical and commercial matters such as contract price and delivery elsewhere in the contract documents.

The following are key items to be completed:

- Date of the contract agreement
- Name and address of both the purchaser and contractor to the contract agreement
- A brief description of work under the contract
- A schedule of contract documents deemed to form part of the contract agreement each identified by title and reference by an exhibit mark
- Signatures of the parties "under hand" or "under seal" if appropriate, and signatures of witnesses.

A sample contract agreement form for a major equipment procurement is shown in Part 5, Section 15.1.5. It is strongly suggested that reference be made to the order of precedence of the listed contract documents should a dispute arise concerning conflicting requirements between any two or more of the contract documents.

It is usual practice to prepare duplicate copies of the contract documents for signature, one for the purchaser and the other for the contractor. Where possible, all documents should be bound in a single volume and in such a manner that a page removal and replacement would be difficult.

Both copies of the contract agreement should be signed by the authorized representatives of the purchaser and the contractor. Although time consuming, all revisions or amendments should be individually initialed by the persons signing on behalf of the purchaser and the contractor.

12.8   THE CONTRACT

The engineer having read the book to this stage may now question: "Is all this contract formation work really necessary in practice?"

Briefly, the contract will evidence:

- The parties who signed the contract
- The transaction covering equipment and services
- Price, method, and terms of payment
- Completion dates in the contract program

Apart from the above, the essential purpose of a contract is to prevent problems arising between the contractor and the purchaser during the course of the work. A well drafted contract should minimize actual or potential problems in key areas such as:

- Avoiding misunderstandings by defining the rights, duties, and responsibilities of the purchaser and the contractor, and agreement on what action will be taken by either the party if various eventualities arise during the course of the contract
- Providing a means of settling disputes through mediation and arbitration
- Providing protection and remedies against default by either the contractor or the purchaser
- Providing protection against unforeseen disasters and other events by covering risk of loss areas through insurance and force majeure (acts of God) provisions

- Avoiding disagreements over performance by defining significant events in the contract program, including stages of contract completion, taking over the equipment, and procedures covering certification for authorizing payments
- Handling contract variations in price and time

In summary, the contract is enforceable at law. It is a formal agreement between the purchaser and the contractor, whereby each accepts to perform certain duties and undertake responsibilities and receives certain benefits. Failure in any way to carry out a duty or responsibility usually involves some form of sanction, either expressed or implied in the contract, or under the law of the land. It is of utmost importance, therefore, that all who are associated with executing a contract are fully aware of their involvements and the precise implications of failing to implement and fulfill them. The liabilities for breach of contract can be heavy and drastic, and can easily exceed the value of the contract itself.

XYZ CORPORATION
2547 TENTH STREET
ANYTOWN NJ 00001

DEVIN RIVER MINE PROJECT

PROJECT No 1712

ENGINEERING SPECIFICATION
No   SP–1203

FIRE WATER PUMPS

CONTRACT No   C–1363

Sheet 1 of 10

ISSUE	DATE	ISSUE DESCRIPTION	BY	APPROVALS	
1	Apr 1 '92	Revised as noted & issued for purchase	G.I.	TWM	
0	Feb 22 '92	Issued for bids	Innes	TWM	

FIGURE 12–1   Example of a specification title sheet revised for purchase.

XYZ CORPORATION	ENGINEERING SPECIFICATION	SHI |TWM|
	FIRE WATER PUMPS	Sheet 2 of 10
	Specification No. SP-1203	Contract No. C/1363	Revision No. 1

5. WORK EXCLUDED

5.1   The following will be furnished by the Purchaser:

(a)   Erection of the pump sets
(b)   Foundations and anchor bolts
(c)   Piping beyond the Contractor's termination points
(d)   Electric wiring from start-up panel to remote initiating and alarm system and between the start-up panel and motor starter
(e)   Electric wiring to motor pump unit and starting equipment

 (f)   Power supply to battery charger and water jacket heater

6. PUMP SELECTION CRITERIA

6.1   The pumps shall be identical having the same rating. Both pump units may be run in parallel and pump characteristic curves shall be such pumps take equal loads.

6.2   The pumps shall be capable of specified rated discharge capacities and heads as listed in the pump data sheets. The Contractor shall select a pump with the operating point close to the peak efficiency on the pump performance curve. Pump speed shall not exceed 1800 rpm. Pump units shall not overload the drivers at any point along the head capacity curve from shutoff head to the maximum capacity for the impeller provided.

6.3   At the rated condition noted in the pump data sheet, the impeller shall be sized for not more than 90 percent of full diameter.

6.4   The pump shutoff head shall not exceed 120 percent of the total rated head.

7. PUMP DESIGN AND CONSTRUCTION

7.1   Pumps shall ~~preferably~~ be the horizontal split, single stage, double suction, closed impeller, volute type. Pumps shall be the Manufacturer's standard approved fire pump design.

7.2   A ~~Richardson~~ Thomas All Metal-Flexible type coupling ~~or equal~~ shall be provided between each pump and driver and shall be designed with a 1.5 service factor based on the driver power noted on the nameplate. Suitable guards and protection shall be provided complying with all prevailing safety standards and codes.

FIGURE 12–2   Example of a specification continuation sheet revised for purchase.

1	Project: Devin River Mine		Area: No. 1 Treatment Plant				Service: Fire water	
2	Manufacturer WORTHINGTON		Size / Type 8LN – 21e			Serial No. +.61a-c		
3	Type Of Driver: Elecric motor		Furnished By: Pump supplier		Mounted By: Pump supplier			
4	Pump Specification No. SP-1203		Pump General Arrangement Dwg No 8LN - 125-G/2					
5	Other Applicable Specifications: SSP-0201, SSP-5101							

6	OPERATING CONDITIONS				PERFORMANCE			
7	Liquid Pumped		tank water		Performance Curve No. ER-4.005			
8	Capacity Flow	Normal	Rated	227	L / s	Speed 1775 rpm NPSH Reqd 3.5 m		
9	Liquid Temp	Normal	Rated	15	° C	Efficiency 80 % kW Rated 323		
10	Discharge Press	Normal	Rated	1150	kPag	Max kW Rated Impeller 410		
11	Suction Press	Max	Rated	12	kPag	Max Head Rated Impeller 136 m		
12	Differential Press	1138	Kpa	116.1	m	Rotation: from Coupling End – ⊂cw⊃ ccw		
13	Specific Gravity at Rated Conditions			1.0		SHOP INSPECTION AND TESTS		
14	NPSH Available			11.2	m	"x" where applicable	Certified	Witnessed
15	Vapor Press at Rated Conditions			1.7	kPaA	Perform w/o Driver		
16	Viscosity at Rated Conditions				Pa • s	Perform with Driver	x	
17	Corrosion / Erosion Caused By: nil					Hydrostatic Test	x	
18	CONSTRUCTION					NPSH Test		
19	Nozzles	Size	Rating	Facing	Location	Shop Inspection	x	
20	Suction	350	ANSI 250	FLAT	SIDE	Dismantle & Inspect		
21	Discharge	200	ANSI 250	FLAT	SIDE	After Test		
22	Case – Mounting ~~Centerline~~ (Foot) ~~Bracket~~							
23	Split (Axial) ~~Radial~~					MATERIALS		
23	Type Volute ~~Single~~ / (Double) ~~Diffuser~~					Pump Case	CAST IRON	
24	Hydrostatic Pressure	3105 kPag				Case Wear Rings	BRONZE	
25	Connection (Vent) (Drain) (Gage)					Impeller	BRONZE	
26	Impeller Diameter Rated 500 / Max 558 / Min 483 mm					Impeller Wear Rings	BRONZE	
27	Mounted (Between Bearings) ~~Overhung~~					Shaft	STEEL	
28	Bearings – Type Radial BALL Thrust BALL					Sleeve (Packed)	BRONZE	
29	Lubrication ~~Ring Oil~~ / ~~Flood~~ / ~~Flinger~~ / ~~Press~~ / (Grease)					Sleeve (Seal)	–	
30	Coupling Mfr THOMAS Model ALL METAL - FLEXIBLE					Throat Bushing	BRONZE	
31	Driver Half Mounted By: (Pump Mfr) ~~Driver Mfr~~ / ~~Purchaser~~					Gland	BRONZE	
32	Packing Type GRAPHITED 15.9 No. of Rings 5 + 5					Lantern Ring	BRONZE	
33	Mechanical Seal Mfr & Model / Type n/a					Base Plate FABRICATED STEEL		
34	Mfr Std No. n/a					MASS OF EQUIPMENT		
35	Base Plate: Common with Driver (yes) no; Drip Rim (yes) no					Pump Net 198 / Pump & Driver 1880 kg		
36	DRIVER							
37		Electric Motor		Steam Turbine		Engine	Gear	
38	Manufacturer	US ELECTRIC						
39	kW / rpm	450 / 1775						
40	Spec No.	SSP-3302						
41	Data Sheet No.	DS-3305-2						
42								

R							
E	1	Mar 29 '92	Bidder's information added and issued for purchase	JCC	TWM		
V	0	Feb 22 '92	Issued for bids – Bidder to complete this sheet	JCC	TWM		
	No.	Date	ISSUE DESCRIPTION	By	App'd	Date	

XYZ CORPORATION	HORIZONTAL CENTRIFUGAL PUMP EQUIPMENT DATA SHEET Equipment Tag No. 1201 Contract No. C-1363	Document No. DS-1205-2A Revision No. 1

FIGURE 12–3   Centrifugal pump data sheet issued for purchase.

XYZ CORPORATION

TECHNICAL REQUIREMENTS FORM

TITLE Fire Water Pumps	**CONTRACT No** C-1363
	PROJECT No 1712
DELIVER TO XYZ Mining Company	
Devin River Mine Project Site	**REVISION No** 1
Cliftin Hill, OHIO 00045	
	SHEET No 1 **OF** 1

ATTACHMENTS

	THE FOLLOWING DOCUMENTS FORM PART OF THE TECHNICAL SPECIFICATION	REV
1	Drawing and Technical Data Submittal Requirements Form	0
2	Fire Water Pump Specification SP – 1203	1
3	Standard General Data and Requirements Specification SSP – 0201	3
4	Standard Electric Motor Specification SSP – 3302	2
5	Standard Noise Specification SSP – 5101	5
6	Pump Data Sheets DS – 1205 – 2A and 2B	1
7	Induction Motor Data Sheet DS – 3305 – 2	1
8	Diesel Engine Data Sheet DS – 7002 – 1	1
9	Drawing No 5-1363-11	1
10	Exhibit "A"	

REV No	ITEM No	QTY & UNIT	DESCRIPTION	ACCOUNT No
	1	1	Fire water pump with electric motor driver and	
			all accessories as specified	as tag
			Tag Equipment: 1201	number
	2	1	Fire water pump with diesel engine driver and	
			all accessories as specified	as tag
			Tag Equipment: 1202	number
1	Apr 2'92		Revised as noted and issued for purchase	G.T.	TWM	
0	Feb 24'92		Issued for bids	G. Tate	TWM	
REV	**DATE**		**ISSUE DESCRIPTION**	**BY**	**APPROVALS**

FIGURE 12–4 Technical requirements form revised and issued for purchase.

PART THREE

THE ROLE OF SECURITIES
IN THE PROCUREMENT

This is the last part of the text dealing with the principles and procedures for the formation of a procurement contract. The previous chapters described guidelines on competitive bidding procedures that culminate with the award of a contract based on the results of a comprehensive bid evaluation. To reach this stage, the purchaser will have committed much valuable time and effort to select a bid and a contractor to execute the work. The purchaser's interests must then be protected, in the first place in that the successful bidder will not decline to accept the contract, and secondly, that the contractor will not fail to complete the work because of incapacity, inefficiency, or some other cause. Hence the need for these securities.

13

BID AND CONTRACT SECURITIES

13.0 INTRODUCTION

When planning a major procurement, it is important to provide for protecting the purchaser's interests from unexpected eventualities that may arise before and after award of contract such as:

- The selected successful bidder declines to enter into a contract to perform the work under the contract, or withdraws the offer prior to notice of award or letter of acceptance
- During the course of the contract, the contractor fails to comply with the terms and conditions of the contractual obligation

These are complex issues and their legal implications are not always fully understood. Nevertheless, the person responsible for formulating the procurement must decide what forms of security are required and the extent of their cover to protect the purchaser's interests.

This chapter describes how these risks can be protected by a provision in the bidding documents of a guarantee in the form of a specified security. The provision(s) must be carefully vetted by a person qualified and experienced in such matters to obtain the protection or indemnification expected by the purchaser in the event of a default by the bidder or the contractor.

13.1 FORMS OF SECURITIES

Securities are issued in the form of bonds (not to be confused with investment bonds). A definition of a bond is the written obligation of one party (the contractor) to answer to another party's (the purchaser) demand should the first party default. It is a certificate or document showing evidence of a debt on which the party that is issuing the bond promises to pay the holder of the bond (the purchaser) a specified amount of money as a pledge of good faith. In the event of these securities having to be provided by the bidders and/or by the successful contractor, the purchaser's requirements must be clearly outlined in the bidding documents. However, it should be noted that bidders/contractors will take account of the costs associated with securities and include them in their bid price.

Specific forms of securities are:

- Cash or equivalent
- On demand bond or bank guarantee
- Surety or conditional bond

Cash or Equivalent

The payment of *cash or equivalent* (certified check) as a security deposit has a disadvantage to both the bidder/contractor and the purchaser. It not only ties up the bidder's and contractor's working capital but also requires the purchaser to account for the funds including interest accrued on the deposited funds when the deposit is returned. Because of this, the preferred method of providing a bid and/or performance security is in the form of a demand or surety bond.

Demand Bond

A *Demand Bond,* also known as a *Bank Guarantee,* is an unconditional undertaking by a bank (and in certain circumstances, by surety companies—see conditional bonds) that it will pay at any time any sum or sums to the maximum aggregate of the amount specified on the bond, on first demand by the purchaser. The purchaser is not required to produce evidence of any loss, or of the bidder's or contractor's default, or of anything else, to support the claim. The advantage of this type of bond to the purchaser especially after award of contract is that he will avoid becoming involved with a possible lengthy litigation to prove his case should the contractor deny a default.

The bond is in the form of a promissory note or irrevocable letter of credit. The cover under the bond is usually an amount of:

- Two percent of the bid price for a bid security, and
- Up to a maximum of 15 percent of the contract price for a performance security

Surety or Conditional Bond

A *Surety or Conditional Bond,* also known as a *Default Bond,* is a bond that is executed by both the bidder, later the contractor, referred to as the "Principal," and the "Surety" (an institution other than a bank, usually an insurance company) whereby the latter guarantees the responsibility for making good to the purchaser, referred to as the "Obligee," any financial loss sustained through the principal's failure to fulfill the specified obligations defined in the bond. In more explicit words, the bond is simply an undertaking by a third party, the surety, to indemnify the purchaser in the event of a default by the bidder or contractor who commissioned the bond. It does not matter what reason caused the default, may it be for incompetence, dishonesty, insolvency, or anything else.

The major difference between a surety bond and a bank's letter of credit is that surety bond issuers become involved in the claims management process, and also assist in the operations to help expedite completion of the contract. The banks usually do not have this specialist expertise to deal with the default problems as is the case with the surety companies.

13.2 SPECIFIC TYPES OF SURETY BONDS

There are numerous forms of surety bonds. In the United States, all government and municipal contracts are required by law to have contract bonds in place to provide assurance that the requirements of the contract will be met. The use of these bonds is also the accepted practice in private industries.

The U.S. contract bonding system includes:

- Bid bonds
- Performance bonds
- Payment bonds
- Maintenance Bonds

Bid Bonds

The purpose of the *Bid Bond* is to guarantee to the purchaser (as "Obligee") that the named contractor (as "Principal") whose bid has been accepted, will enter into a formal contract with the obligee within the time required and provide any specified performance and payment bonds within a specified time (usually 10 days). In other words, the bid bond could be said to ensure the good faith of the contractor in submitting the bid. Should he fail to accept the contract as awarded and to supply the required above mentioned bonds, the surety is then obliged to pay to the obligee, up to the penal sum of the bid bond, the difference between the original accepted bid, and the contract price for which the obligee may legally sign a contract with another party to perform the work.

A bid bond penalty is normally 5 to 20 percent. For larger projects, however, the bid bond penalty should be kept in line with the out-of-pocket loss to the purchaser (for example, cost of reletting the contract) should the bidder whose bid was accepted default in entering into a contract with the purchaser. Care must be taken that this upper limit is adequate.

The bid bond lapses when the purchaser confirms acceptance of the contract by signing the award of contract or letter of acceptance, or when the contract is awarded to some other party.

Returning bid bonds to unsuccessful bidders is unnecessary as the bond automatically becomes void after a contract is signed.

Instead of a bid bond to accompany each bid, a cash or equivalent is normally acceptable as previously discussed. The cash deposit is returned to all bidders when the purchaser has accepted a bid, or at the end of the validity period, whichever comes first.

An example of a bid bond is shown in Figure 13–1 (page 13–8).

Performance Bonds

The basic purpose of providing a *Performance Bond* is to make certain that, in the event of the contractor's ("Principal") failing to comply with the terms and conditions of the contractual undertaking, the purchaser ("Obligee") will be indemnified, or be protected by the surety against loss suffered by reason of the contractor's default up to the amount of the bond penalty, or the surety will arrange completion of the contract. The performance bond, up to its monetary limit, ensures that the contractor will punctually, and faithfully, observe all his obligations under the contract. Performance not only covers the technical completion of the work but also nonperformance as a result of bankruptcy, and commercial takeover.

The bond penalty is often equal to 100 percent of the contract price and is favored by the purchaser since it relieves much of the risk arising from default by the contractor.

Failure of performance by the contractor must be declared in writing by the purchaser to the contractor and to the surety. Depending on the surety's terms of commitment, the surety must in some manner:

- Itself complete the contract, *or*
- Arrange completion of the work by another contractor, and pay the purchaser any extra cost to him for the balance of the work to be completed, *or*
- Elect to pay the purchaser up to the full amount of penalty of the bond
- Provide financing so the existing contractor can complete the work

A default by a contractor generally creates numerous difficulties for the purchaser. The defaulting contractor will seldom have any interest to make things easier for the new contractor assigned to complete the work. The purchaser would be wise to make sure that the successor is capable, reliable, and financially able to complete the contract.

An example of a performance bond is shown in Figure 13–2 (page 13–9).

Payment Bonds

A *Payment Bond,* also called *Labor and Materials Payment Bond,* to be furnished by the contractor ("Principal") is a guard against possible difficulties arising from his failure to pay for labor and materials furnished for use on the project. The bond protects the purchaser ("Obligee") against loss should the contractor fail to make payments to all claimants such as subcontractors and other suppliers engaged on work under the contract. Under the bond, a claimant is one who has direct contact with the contractor for the provision of labor and/or materials required for use under the contract.

It should be noted that subcontractors to the contractor's subcontractors usually are not claimants within the terms of a payment bond. The United States government requires payment bonds by statute to embrace the suppliers and subcontractors to be covered as claimants so bond forms as well as statutes must be reviewed to see how far coverage extends. A subcontractor may have already entered into a bond to protect his interest to ensure a supplier's financial obligation to make payments for accounts.

The need for protection under this bond is currently becoming more apparent because of the increasing consequences resulting from laws of liability. Like the performance bond, but unlike the bid bond, the payment bond is generally written to cover 100 percent of the contract price.

Maintenance Bonds

The function of a *Maintenance Bond* is to ensure the purchaser ("Obligee") that the contractor ("Principal") fulfills his obligation under the labor and material equipment warranty provision in the contract. A maintenance bond is usually required where no performance bond was issued on the contract, or for a particular warranty of the work under the contract that extends beyond the specified warranty provision. The maintenance bond for the entire equipment is usually written to cover 15 to 20 percent of contract price or the value of the work being guaranteed.

13.3   PURCHASER'S OBLIGATIONS AND CHANGES TO THE CONTRACT

The purchaser ("Obligee") to whom the bond is payable on a default on the part of the bidder or contractor ("Principal") is expected to adhere to the terms of the bond as well as the terms and conditions of the contract in order to preserve his protection or the surety will declare the bond void. Should there be any change during the course of the contract, the purchaser must ensure that such changes do not invalidate the bond. Any material change must be brought to the attention of the surety since the contract is tied to the bond.

The surety will determine the effect upon its risk. Material changes to the contract may include:

- Change in scope of work
- Additional work involving an increase in the contract price
- A change in the type and nature of the work
- A change to the contract program involving a change of the time for completion of the work

13.4   CLAIMS WHERE THE CONTRACTOR DENIES DEFAULT

A claim under a bond where a solvent principal denies he is in default places both the obligee and the surety in a serious predicament. In such a situation, it may left to a court or arbitration panel to determine the validity of the default and the course of action to be taken by the surety.

13.5   BASIC BOND CHECKLIST

There are numerous other forms of surety bonds. A number of technical organizations and finance institutions have developed model bond forms specially tailored to meet their respective requirements. Various government authorities at all levels require their own particular format, and statutory wording as prescribed by law and regulations under law. While the purchaser's engineer assigned to a contract should have sufficient understanding on issues relating to bonds, a competent legal advisor should always be consulted when dealing with bond matters. Of main concern is the validity and adequacy of the terms of the bond to meet the requirements of the particular procurement undertaken.

To assist the purchaser's engineer regarding bonds, the following is a basic checklist of information a surety bond should contain:

- The legal name and business address of the principal (contractor), and state of incorporation—who is bonded
- The name(s) and business address of the surety(ies) who issues the bond
- The name and address of the obligee (purchaser)—to whom the bond is payable
- The sum of the bond
- Date of bid submission (for bid bond) or date of contract

- Identification of equipment covered under the contract
- The conditions of the obligation under the bond
- Evidence of legal authority of person signing the bond on behalf of the surety (usually power of attorney)
- Time limits applying to bonds

BID BOND

Bond No.   Amount $

Know All Men By These Presents,

That we,

                                                    (hereinafter called the Principal),

of

and   , a corporation duly organized under

the laws of   , as Surety, are held and firmly bound

unto   , hereinafter called the Obligee,

in the sum of   Dollars ($)

for the payment of which we, the said Principal and the said Surety, bind ourselves; our heirs; executors; successors and assigns, jointly and severally firmly by these presents.

Sealed with our seals and dated this   day of,   , 19   .

WHEREAS, the Principal has submitted a bid dated

for

NOW, THEREFORE, THE CONDITION OF THIS BOND IS SUCH, that if the Obligee shall accept the bid of the Principal and the Principal shall enter into a contract with the Obligee in accordance with the terms and conditions of such bid, and give such bond or bonds as may be specified in the Bidding Documents with good and sufficient surety for the faithful performance of such contract, or in the event of the failure of the Principal to enter into such contract and give such bond or bonds, if the Principal shall pay to the Obligee the difference, not to exceed the penalty hereof, between the amount specified in the said bid, and such larger amount for which the Obligee may legally and in good faith contract with another party to perform the work covered under the said bid, then this obligation is null and void, otherwise remain in full force and effect.

_____
                                                        Principal

By: _____

By: _____

FIGURE 13–1   Sample form of bid bond.

PERFORMANCE BOND

Bond No.   Amount $

Know All Men By These Presents,

That we,

                                                          (hereinafter called the Principal),
as Principal, and the   a corporation duly organized under
the laws of the   , (hereinafter called the Surety), as Surety, are held and firmly bound unto

                                                          (hereinafter called the Obligee),
in the sum of
Dollars
($), for the payment of which we, the said Principal and the said Surety, bind
ourselves, our heirs, executors, administrators, successors and assigns, jointly and severally,
firmly be these presents.

Sealed with our seals and dated this   day of,   ,   19   .

WHEREAS, the Principal entered into a certain Contract with the Obligee, dated   , 19   ,
for

in accordance with the terms and conditions of said Contract, which is hereby referred to and
made a part hereof as if fully set forth herein.

NOW, THEREFORE, THE CONDITION OF THIS OBLIGATION IS SUCH, that if the above
bounden Principal shall well and truly keep, do and perform each and every, all and singular,
the matters and things in said Contract set forth and specified to be by said Principal kept,
done and performed, at the times and in the manner in said Contract specified, or shall pay
over, make good and reimburse to the above named Obligee, all loss and damage which
said Obligee may sustain by reason of failure or default on the part of said Principal so to do,
then this obligation shall be null and void; otherwise shall remain in full force and effect, subject, however to the following conditions:

Page 1 of 2

FIGURE 13–2   Sample form of performance bond.

PERFORMANCE BOND (cont)

Any suit under this bond must be instituted before the expiration of two (2) years from the date on which final payment under the Contract falls due.

No right of action shall accrue on this bond to or for the use of any person or corporation other than the Obligee named herein or the heirs, executors, administrators, or successors of the Obligee.

_____
Principal

By: _____

By: _____

Page 2 of 2

FIGURE 13–2   (Continued)

PART FOUR

SPECIFICATION WRITING AND PRODUCTION TECHNIQUES

One of the most important skills an engineer must develop, apart from technical know-how, is that of using language to its best effect so meaning and intent are always clear. The objective when writing specifications for procurement contracts is to convey the precise requirements not only to the contractor but to all who are associated with the contract. The means required to achieve this objective are the effective use of simple, precise language and the ability to produce a document that is clearly set out and therefore easily read.

This chapter examines the process of writing and producing effective specifications. It deals with issues such as the use of language to convey meaning, the development of format, style, and document presentation, and techniques used for the production of specifications.

14

SPECIFICATION STYLE AND DOCUMENT FORM

14.0 INTRODUCTION

A specification writer must possess a complete understanding and knowledge of the work to be accomplished for a particular contract. As specifications are written instructions setting forth the complete conditions and requirements of the work, it must be assumed that this person not only possesses adequate knowledge of the requirements but also the ability to write clearly when expressing thoughts and ideas. The specification writer's work will be of little value unless the ideas are clearly understood by those who are later responsible for executing the work. Some specification writers tend to include complex legal terminology or confusing instructions in specifications. Such practices have resulted in misinterpretation of general requirements.

The skill of writing successful specifications is usually developed over time, or with continuous efforts being made for improving clarity and presentation. A common practice is to use an old specification from a previous contract because it covers a similar scope of requirements, and then the necessary changes are made by deleting or adding text and data to suit the new project. This can be a dangerous practice. The old specification may contain errors that may have only been discovered during the course of the work, the specification itself may not have been revised. The end result is that the new document contains a repeat of the same errors.

14.1 SPECIFICATION-WRITING TECHNIQUES

A specification writer should bear in mind that every part of a specification carries along with it a monetary price, be it for manhours expended for engineering design, cost of material, cost for fabrication, or any other requirement that a contractor has to perform under a contract. A vaguely worded specification usually infers that the specification writer is unsure of what is actually wanted. Statements such as "additional drawings may be required by the Engineer" and "testing requirements will be at the discretion of the

Engineer" make bidding for a contract extremely difficult as a bid reflects every statement in the specification. Vague specification provisions may result in expense for the purchaser because bidders will have covered all likely contingencies in their bid price.

The specification writer should put himself or herself in the place of the specification user and determine whether the instructions are easily understood. If not, the user will not be able to execute them.

A good specification is concise, using as few words as possible to completely describe the requirements. Complicated phrases should be avoided, and short, simple words should be used. Verbosity and repetition lead to ambiguity.

The following sections outline the techniques for writing clear, concise, effective specifications.

14.1.1   Contradictions, Grammar, and Spelling

Contradictions

Extreme care should be taken to avoid contradictions:

- Within the specification
- Between the (detail) specification and a general specification
- Between the specification and the bid
- Between the specification and the drawings
- Between the specification and the contract conditions

Contradictions within a contract must be eliminated because they may lead to invalidation of the contract, and will certainly lead to costly and time-consuming disagreement between the parties.

If a standard specification is used in connection with the detail specification, the detail specification should contain a paragraph as follows:

> The requirements of the detail specification shall govern in case of conflict between the standard specification and the design and performance specification.

Grammar

Pronouns. It is best to avoid the use of pronouns in specification writing and to simply repeat the nouns. An example of the confusion that can arise from the use of a pronoun is, "Contractor shall advise the Purchaser after inspecting the surface of a casting for any defects that might endanger his work." The word "his" can appear to apply to the

"Purchaser," although obviously the writer is referring to the "Contractor." If this matter ever came to court, as it might, syntax would win over synesis every time. Legal entities such as "Contractor" and "Purchaser" cannot be referred to by personal pronouns such as "his" and "their."

The only pronoun usage permitted in technical writing is that of the impersonal pronoun "it," but even "it" must be used with care, for example, "Contractor shall furnish all manpower that it requires."

Shall and Will. For the purposes of specification and contract writing, the following usage governs:

Shall is mandatory. Whatever the contractor is required to do must be expressed by the use of the auxiliary "shall," for example, "Contractor shall furnish six sets of shop drawings within ten days after award of a contract for approval by the Purchaser."

Will is not mandatory. Will is merely an expression of an intended future action, for example, "Electrical work will be done by others." Use "will" in connection with anything to be done by the Purchaser.

Spelling. Most annoying to a reader is incorrect or inconsistent spelling. When in doubt, spelling checks should be made by referring to the latest edition of the dictionary selected, such as *Webster's Dictionary* as a guide to American spelling, or the *Concise Oxford Dictionary* for British spelling. There are many instances where American spelling differs from the British form. Examples include:

British	*American*
catalogue	catalog
centre	center
cheque	check
calibre	caliber
colour	color
favour	favor
fulfil	fulfill
labour	labor
litre	liter
metre	meter (measurement of length, not measuring apparatus)
skilful	skillful
sulphur	sulfur

Another group of words are those that end with only *-se* in British style but end with only *-ze* in American style.

analyse	analyze
organise	organize
specialise	specialize

Abbreviations. These are dealt with in Appendix D.

14.1.2   The Do's and Don'ts

The following have particular relevance to contract documentation:

The Do's

- Ensure that specifications are written by engineers possessing the necessary writing skills and who have an adequate understanding of all conditions and requirements of the subject matter
- Be sure the specification is clear and easily understood
- Ensure that the contents of the specification are fair. Unfair provisions by the purchaser will inevitably cause complaints from the contractors
- Be definite and precise with descriptions and requirements
- Give directions and avoid making suggestions
- Use correct English grammar and choose words carefully, so their meaning is unambiguous
- Limit each section to one paragraph, covering the subject matter completely in that paragraph
- Ensure that the specification agrees with the referenced equipment data sheets, drawings, and all applicable codes, standards, statutory rules, and regulations
- Be sure that all requirements in the specification can be complied with and met, and that the scope of work can be supplied by equipment manufacturers
- Note that federal, state, and provincial laws applying to a contract have precedence over whatever the contract may say

The Don'ts

- Do not attempt to place all the risks upon the contractor
- Do not use long sentences

- Avoid the use of multisyllabic or harsh words
- Do not include incorrect information that may eventually require later revision
- Do not present jumbled information that would result in someone's having difficulty in understanding or finding needed information
- Avoid omissions
- Avoid abbreviations; spell out if in doubt
- Do not provide vague description of requirements, or poor definitions of the scope of work or division of responsibilities
- Do not specify any requirement that cannot be enforced
- Avoid ambiguity
- Do not make contradictions or contradictory statements
- Do not make reference to out-of-date documents, or superseded technology
- Do not make reference to nonexisting data in the specification
- Avoid making unnecessary duplications
- Avoid making indefinite statements such as "acceptable to the Purchaser"
- Do not call for requirements that are not in the specification's scope of work
- Avoid gratuitous instructions to the contractor
- Do not include commercial matters in the technical specification, such as liquidated damages, prices, currency exchange rates, customs duties, and delivery requirements

14.2 DOCUMENT PRODUCTION TECHNIQUES

Preparation of the purchaser's bidding and contract documents for a procurement invariably is only done once. Publication of these documents is generally accomplished using company personnel and word processing or desktop publishing software. The following is a guide for editing and publishing the documents.

14.2.1 Word Processing

The use of word processing or desktop publishing software has now become the predominant method of producing specifications. The simple typewriter has become antiquated because data need only be entered once into a computer using readily available word processing software. Computerized word processing avoids having to type draft copies, possibly retype them where changes are needed, and then retype the final copies for formal presentation. Once the specifications have been entered and saved on disks, errors, additions, modifications, editing, and making other changes to any part or format of the text can be easily accomplished without the need to retype. Most word processing software offers a user-friendly interface to get the job done in the minimum amount of time without a steep learning curve. With use of a mouse (a handheld device that when moved across a surface, moves a pointer on your screen), one can access formatting and style options, such as "cut and paste," fonts, page dimensions, and so on, by simply clicking a button on the mouse. Tables, charts, and graphics can all be added if required. Clear and highly legible hard copies can then be produced on modern, high-speed laser printers.

14.2.2   Paragraphs, Tables, and Illustrations

Each page of a specification should be identifiable to safeguard against loss, to ensure a page is located in the correct document when other similar documents are handled, that pages appear in their correct sequence, and for countless other reasons.

Identification techniques vary considerably, but the information should always provide clear and swift page identification. Methods include a header or footer on each page showing:

- Purchaser's company name
- Document title
- Document number
- Contract number
- Document revision number
- Page number

Tables and Illustrations

Tables and illustrations such as flow diagrams, schematics, and block diagrams are essential tools for assisting bidders and contractors to understand the purchaser's precise requirements. Nothing can be more frustrating than having to plough through countless pages of highly complex material trying to form a mental picture of what the purchaser is describing. Flow and block diagrams and similar figures, if not able to be placed on the page vertically (portrait), should be readable horizontally (landscape) when the document is rotated 90 degrees clockwise.

To avoid difficulties in finding cross-referenced tables and illustrations, the material should be located as near as possible to the point where referenced in the text and should always be placed *after* its text reference. Full-page illustrations should be referenced in the text as with any other illustrative material, but larger-sized illustrations in the form of fold-out pages should preferably be presented as an appendix and located at the rear of the document.

14.2.3   Appendices and Specification Parts

Appendices are normally created to include a series of large or complex drawings, reports, schedules, and other supplementary material that are presented separately from the main text of the specification.

Specification parts are separate volumes of the contract such as the technical section, commercial section, and drawings. They can be provided separately to bidders, but only on request.

14.2.4   Covers and Binders

An imaginative and well executed cover on bidding documents always creates a very positive image of the caliber of the purchaser's company (and also bidders submitting their bids). Covers should be of adequate weight to survive repeated handling.

The preferred methods of binding documents are plastic combs. Two- or three-hole bindings with D-rings (or filing type) devices should be avoided as pages can easily be lost or torn out. Comb-bound documents may be more than 25 mm thick, and when opened, allow the facing pages to lie flat. Pages can easily be removed and replaced if necessary but are less likely to become adrift. This method of binding can be accomplished with inexpensive in-house equipment.

PART FIVE

DOCUMENTS FOR A SAMPLE PROCUREMENT

This final part of the text is designed to show how each of the key documents for a major equipment procurement has its bearing upon the contract formation process.

The documents have been prepared for a sample procurement covering the principal operations from the initial invitation to bid through to the final preparation of the contract documents for a particular procurement. The aim is to demonstrate in a practical manner, the general features, structures and methodology that are needed for competitive bidding, equipment selection, and awarding the contract.

SAMPLE DOCUMENTS

15

INTRODUCTION TO
THE SAMPLE PROCUREMENT

The importance of this part of the book lies intrinsically in the contents of the documents and their presentation for a major equipment procurement through competitive bidding. It is not intended to limit or constrain any procedures made necessary by unique requirements, or special circumstances not addressed in this procurement example. The documentation in all cases should be prepared to meet the specific requirements of a particular procurement.

This particular example covers the supply and delivery of high capacity centrifugal air compressor units of proven design for a continuous process operation, and is equipment that most engineering disciplines would be familiar with. To present procurement methodology for more complex equipment could possibly lead to confusion when describing the formulation of the bidding and contractual documents. The names of the purchaser, the project, the procurement, the personnel referred to as being responsible for compiling the documents, and the successful bidder are entirely fictitious. Although some of the documents and requirements would seem unnecessary for air compressors, the sample procurement has been made deliberately complex so that it can illustrate general procurement practices where the following elements are involved:

- Significant financial expenditure
- Major items of equipment required for the purchaser's plant
- The procurement of equipment requiring the preparation of complex commercial and technical documentation for bidding, and the placing of a formal contract as compared with simple "off-the-shelf" materials

The purchaser's program for the procurement of the compressors is shown on page 15–2. To illustrate certain procedures in the procurement formation process, it is assumed that the initial release of bidding documents called for the supply of two compressor units. However, during the bidding period, the purchaser found that production capacity had to be upgraded with the result that three units were required immediately. This necessitated forwarding a supplemental notice to all bidders notifying them of the change but not extending the bidding period.

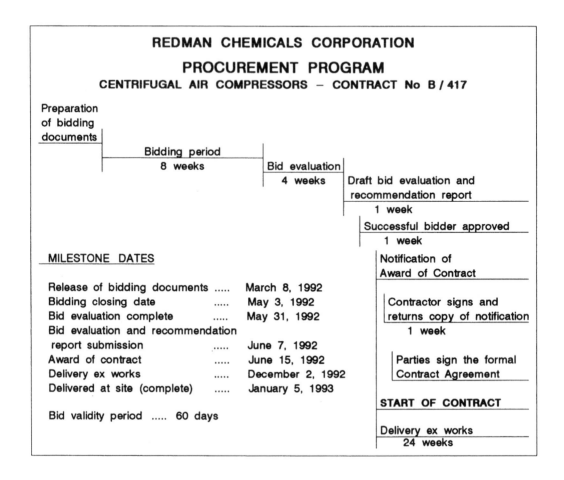

Other items relating to this procurement are:

- Invitations to bid are made worldwide, and reference is therefore made to currency exchange rates, and import duties
- Conditions of the Procurement Contract have not been included since the purchaser may use a standard model form as explained in Chapter 5, or use an in-house produced set of terms and conditions
- Incoterms are used to define equipment loading, transportation, unloading, and insurance responsibilities
- Bid and contract performance securities have not been specified as it is left to the purchaser's discretion whether they would be required
- The contract price is fixed and not subject to price adjustment in order to avoid complex escalation formulae relating to consumer price indexes that vary from one country to another

For this procurement, there were seven bidders on the purchaser's list of bidders who were invited to bid as shown in the *Bidding Status Monitor Form.*

REDMAN CHEMICALS CORPORATION

BIDDING STATUS MONITOR

CENTRIFUGAL AIR COMPRESSORS – CONTRACT B / 417

Programmed date for inviting bids ..... March 8, 1992
Programmed date for receiving bids ..... May 3, 1992

	Name of Bidder Invited to Bid	Date Invitation to Bidder	Date Bid Received	Bid Validity To	Quoted Delivery Date
1	Bidder A				
2	Bidder B				
3	Bidder C				
4	Bidder D				
5	Bidder E				
6	Bidder F				
7	Bidder G				
8					
9					
10					

When bidding closed, Bidder F had declined to bid. Bidder G forwarded a facsimiled offer which was classed a nonresponsive bid and therefore rejected.

Regardless of the size and complexity of the equipment, this procurement example will show how the purchaser has planned and developed the bidding and contract documents with emphasis placed on the technical requirements and the bid evaluation process. The point to be recognized when drafting the technical documents is, "if you don't ask for it, most likely you won't be getting it," since the onus to furnish nonspecified peripheral equipment should not be placed on the contractor.

Furthermore, if the specified scope of supply leaves much to the discretion of the bidders, they well may submit a low price offer meeting "bare-bone" requirements in order to secure the contract. In such instances, there would be no need for a detailed bid evaluation, and the purchaser may eventually be forced to issue formal change orders to the contract (or variations to the contract) for omissions, and pay a large price for his lack of attention at the time of drafting the technical specification. It is therefore at the *front-end* of the contract where much time and money can be lost or saved on the procurement depending on how well the documents have been prepared and how well the bidding process has been administered.

As has been mentioned throughout this book, the documents presented are for guidance purposes only. While it is hoped that the information and analysis of the material shown should lead to a better understanding and application of the procurement formation process, it is important that readers do not rely upon them as being a complete equivalent for their particular project. Nor should any of the sample documents be directly copied, since those for an actual procurement should be carefully verified for scope and content before being used. Only certain sections of the procurement process can be standardized and many readers may prefer to adhere to their in-house method of developing these documents.

15.1   REQUIREMENTS FOR BIDDING

15.1.1   Invitation to Bid

REDMAN CHEMICALS CORPORATION
MONTCLIFF AVENUE
REEDTOWN CA 00012

TELEPHONE: 035 478 925   FACSIMILE: 035 734 045

INVITATION TO BID

March 8, 1992

....
(Bidder's company name)

....
(Address)

....

Attention: ....
Sales Manager

Altona Plant Expansion Development Project No. 7780
Centrifugal Air Compressors - Contract No. B/417

You are invited to bid for the supply of two (2) Centrifugal Air Compressors to be installed at our Altona Plant in accordance with the enclosed three (3) sets of bidding documents.

Please acknowledge receipt of this transmittal by return facsimile confirming that you will be bidding for this contract. Should you decline to bid, please return all documents to the undersigned together with your covering letter.

Sincerely,
Redman Chemicals Corporation

K. T. Kenwood - Project Manager

15.1.2 Instructions to Bidders

REDMAN CHEMICALS CORPORATION
MONTCLIFF AVENUE
REEDTOWN CA 00012

INSTRUCTIONS TO BIDDERS
Centrifugal Air Compressors—Contract No. B/417

The Bidder shall submit his bid with the understanding that these instructions to bidders are intended to cover all of the work to be performed under the contract. Unless expressly excluded in writing, any or all equipment, material and labor not indicated in the bidding documents but which may be necessary to complete any part of the work, are to be furnished by the contractor without extra charge or cost to the Purchaser. Any intended omission(s) must be stated by the Bidder in writing in his bid, otherwise it will be understood that all work is included.

The bid shall be properly executed in every respect. Failure to comply with these Instructions to Bidders may be cause for rejection of the bid at the Purchaser's option.

A   GENERAL

A.1   The Purchaser and address are: Redman Chemicals Corporation
                                        Montcliff Avenue
                                        Reedtown CA 00012

              Telephone Number: 035 478 925
              Facsimile Number: 035 734 045

A.2   The Contract Number is B/417

A.3   The Contract is for:

Furnishing and delivering Centrifugal Air Compressors, together with the services of Advisors for installation and commissioning, all in accordance with Contract B/417

A.4   The type of contract is lump-sum fixed price (not subject to escalation) and a unit price schedule for extra work that may be requested by the Purchaser.

A.5   Delivery of the compressors is to the Purchaser's site at the address above.

A.6   Bidders are to state the best possible delivery to the site in weeks from date of award of contract.

A.7   A bid security is not required for this contract.

B   BIDDING DOCUMENTS

B.1   The equipment and services to be supplied by the Contractor, procedures for bidding and the terms and conditions of Contract No. B/417 are contained in the bidding documents.

INSTRUCTIONS TO BIDDERS (cont)

CENTRIFUGAL AIR COMPRESSORS—CONTRACT No. B/417

The following are the bidding documents:
- (a) The Invitation to Bid
- (b) This Instructions to Bidders
- (c) Content of the Bid
- (d) Bid Form
- (e) General Conditions of the Procurement Contract
- (f) Supplementary Conditions
- (g) The Technical Requirements Form (including all listed attachments)

B.2   Should any Bidder have doubt as to the true meaning of any section or part of the bidding documents, or finds omissions or discrepancies, he may submit in writing a request for clarification and interpretation of the section or part in question up to fourteen calendar days prior to bid closing time.

B.3   A request for additional information should be made to Mr. K. T. Kenwood, Project Manager at the address noted in A.1 above.

B.4   Any amendment or addition made by the Purchaser to the bidding documents during the bidding period will be forwarded to each bidder in the form of a Supplemental Notice. Each supplemental notice becomes part of the bidding documents and will supersede the appropriate portion of the bidding document previously issued. The Bidder is required to acknowledge receipt of such notice.

B.5   The bidding documents remain the property of the Purchaser and shall be used for the specific purpose of preparing and submitting a bid. Upon receiving notice of award of contract to another bidder, unsuccessful bidders shall return the bidding documents to the Purchaser.

C   PREPARATION OF BIDS

C.1   The contents of the Bidder's proposal and all documents exchanged between the Purchaser and Bidder during the bidding period shall be written in the English language. Printed literature furnished by the Bidder as part of the bid and written in another language shall be accompanied by an accurate English translation.

C.2   Bids shall be prepared using the International System of Units (SI).

C.3   Bid prices shall be quoted in U.S. Dollars.

C.4   Bidders shall quote all prices exclusive of sales tax. Appropriate instructions in respect to sales tax will be given by the Purchaser at time of award of contract.

C.5   All royalties or other charges as contained in the Content of the Bid document shall be included in the bid price.

C.6   Bids received by the Purchaser shall remain valid for a period of sixty (60) days after the bid closing date.

C.7   The Bidder's offer shall be in accordance with the Content of the Bid document. Noncompliance features shall be listed in a separate schedule.

INSTRUCTIONS TO BIDDERS (cont)

CENTRIFUGAL AIR COMPRESSORS—CONTRACT No. B/417

C.8 The Bidder may offer an alternate proposal to the one specified in the technical specification provided he considers that the alternate design and construction are equivalent or superior to the requirements of the Purchaser's technical specification. Any such alternative bid submitted shall be clearly marked "Alternative Bid." The Bidder is still obligated to submit a conforming bid in strict compliance with the requirements specified in the bidding documents.

C.9 Partial bids will not be considered.

D SUBMISSION OF BIDS

D.1 All documents submitted by the Bidder should be prepared preferably on letter type pages of 216 x 279 mm page size.

D.2 The Bidder shall submit, typed, the Original bid form together with two (2) Duplicate copies of the bid, each clearly marked "Original" and "Duplicate Copies," dated and signed under hand or under seal. The name and the position of the person(s) signing the form must be typed directly below the signature(s). The Bidder's signature(s) must have a witnessing signature including the position held by the witnessing person.

D.3 Bids are to be lodged in a sealed envelope clearly marked "Proposal" and the Contract No. B/417 and be presented or mailed to:

> Redman Chemicals Corporation
> Montcliff Avenue
> Reedtown CA 00012
>
> Attention: Mr. K. T. Kenwood

D.4 Bids will be received by the Purchaser until closing time and date which is 2:00 PM (Pacific Standard Time) May 3, 1992. Any bids received after this time will be returned unopened to the Bidder.

E RECEIPT OF BIDS

E.1 Receipt of every bid will be acknowledged by facsimile if the bid is mailed or by letter if delivered by hand.

E.2 Bids will not be opened prior to bid closing time.

E.3 Bids will not be opened in public after bid closing time.

E.4 Bids may be withdrawn only by a written or facsimile request received by the Purchaser prior to bid closing time and date. Bids may be modified only by written notice received by the Purchaser prior to such time and date.

INSTRUCTIONS TO BIDDERS (cont)

CENTRIFUGAL AIR COMPRESSORS—CONTRACT No. B/417

F BID EVALUATION

F.1 For the purpose of the comparison bid evaluation, consideration will be given but not limited to the following items:

(a) Qualifications of the Bidder

(b) Price and conformance to contract documents

(c) Time for delivery

(d) Efficiency including power consumption at equipment guaranteed performance

(f) Interchangeability of key compressor components

(g) Location of appropriate facilities adequately staffed for providing prompt availability of component parts and maintenance services

(h) Applicable elements or factors not specifically stipulated or provided herein, which would affect the ultimate cost to the Purchaser.

G AWARD OF CONTRACT

G.1 The lowest priced bid will not necessarily be accepted.

G.2 The Purchaser reserves the unconditional right to reject any or all bids.

G.3 The Purchaser further reserves the unconditional right to consider any non-conformity in bids received when such consideration is in the interest of the Purchaser.

G.4 The Purchaser will issue a formal notification of acceptance to the successful bidder and that the bidder shall be required to enter into a formal Contract Agreement with the Purchaser to execute the contract.

G.5 The successful bidder shall deliver to the Purchaser within seven (7) days after having received the formal Contract Agreement, the two copies of the Contract Agreement for Contract No. B/417 duly executed.

G.6 No advertisement relative to the acceptance of the bid shall be published in any newspaper, magazine, journal, or other advertising medium unless such advertisement has first been approved by the Purchaser.

K. T. Kenwood
Project Manager

15.1.3   Content of the Bid

This section presents an example of a series of schedules to be completed and returned by bidders for a typical procurement such as the centrifugal air compressors. The structure and scope of the schedules for an actual procurement will depend on the particular information sought for bid evaluation purposes. In this sample procurement, it has been assumed that portions of the equipment, if not all, will be secured from overseas sources.

All schedules should have an appropriate heading. To conserve space, the schedules shown are illustrative only. Normally, each schedule may occupy one or more pages.

REDMAN CHEMICALS CORPORATION
MONTCLIFF AVENUE
REEDTOWN CA 00012

CONTENT OF THE BID
Centrifugal Air Compressors—Contract No. B/417

INFORMATION TO BE PROVIDED WITH BID

The Bidder shall provide all information and complete all schedules in accordance with this Content of the Bid.

SCHEDULES—GENERAL

The Bidder shall provide a written statement, dated and signed by an authorized person, for the following:

(a)   That the Bidder's offer incorporates everything for the safe and convenient operation of the equipment whether or not it has been detailed in the bid or is specified in any of the technical documents noted in the technical requirements form

(b)   Confirmation of the Bidder's adherence to the contract program provided in the Supplementary Conditions for this Contract B/417

(c)   Completed Schedule T3 showing clearly and in the order of the relevant clause any proposed departure from the specification(s) including a statement that the offer complies with the specification(s) in all other aspects

(d)   A statement confirming that the Bidder's proposed subcontractors as listed in Schedule T10 are familiar with all aspects of the contract and will comply with relevant requirements thereof

(e)   A statement that the Bidder agrees to ensure the availability of spares and service during the lifetime of the equipment

CONTENT OF THE BID (cont)

CENTRIFUGAL AIR COMPRESSORS—CONTRACT No. B/417

(f)   The name, address and telephone number of the Contractor's Representative available for discussion on the matters relating to this bid:

Name: ....
Address: ....
                                    ....
Telephone Number: ....
Facsimile Number: ....
Name of Bidder's Company: ....

SCHEDULES—COMMERCIAL

SCHEDULE CI — PRICES

The trade terms "Ex Works" and "FOB" shall be defined by the current edition of Incoterms. A copy of the definitions is attached.

		Amount $US	Percentage
(a)	For the Imported Portion:		
	(i) Components FOB	....	....
	(ii) Ocean Freight	....	
	Subtotal Sum		
(b)	For the Domestic Portion:		
	(i) Components Ex Works	....	....
	(ii) Land Freight	....	
	Subtotal Sum		
			100.00 %
	Contract Price		

CONTENT OF THE BID (cont)

CENTRIFUGAL AIR COMPRESSORS—CONTRACT No. B/417

SCHEDULE C2—IMPORTED ITEMS, SEGREGATION OF CONTRACT PRICE

The Bidder shall set out below, the information for dutiable material of overseas manufacture with the exchange rate noted at the date of the bid:

COUNTRY OF ORIGIN	ITEM	FOB VALUE	EXCHANGE RATE	CUSTOMS TARIFF ITEM	CUSTOMS DUTY RATE

SCHEDULE C3 — PRICES FOR SITE ADVISORY SERVICES

Services for a Commissioning Engineer for an estimated fifteen (15) calendar days at $US .... per calendar day.

Transportation, accommodation, and other expenses are reimbursed in accordance with the supplementary conditions for this contract.

SCHEDULE C4 — UNIT RATES FOR ADDITIONAL SERVICES

The Bidder shall complete the schedule of unit rates for the Contractor's personnel to under-take additional services when requested by the Purchaser. Rates shall be quoted for a Senior Engineer, Engineer, Draftsman, Inspector/Expediter, Field Engineer, Installation and Commissioning Engineer.

The rates shall be inclusive of allowances such as insurance, fringe benefits, statutory payments, annual leave, sick leave, provision for long service leave, superannuation, and other welfare expenses.

PERSONNEL CLASSIFICATION	UNIT RATE $US PER HOUR

CONTENT OF THE BID (cont)

CENTRIFUGAL AIR COMPRESSORS—CONTRACT No. B/417

SCHEDULE C5 — TIME FOR DELIVERY

To meet the Purchaser's planning of the Work for this project, the Bidder is requested to submit the earliest delivery date Ex Works and estimated delivery at site in weeks after award of contract.

However, delivery at site of the two compressors shall not exceed January 5, 19__, and any earlier date will be favored by the Purchaser. This shall be based on an award being made no later than June 12, 19__.

(a)   The first compressor complete Ex Works within   ....   weeks
        The first compressor delivered at site within   ....   weeks

(b)   The second compressor complete Ex Works within   ....   weeks
        The second compressor delivered at site within   ....   weeks

SCHEDULE C6 — SHIPPING INFORMATION

The Bidder shall state in the bid:

- Approximate total shipping mass (kg) and the mass per unit
- Approximate mass (kg) of the heaviest piece or package of material
- Approximate number of crates and/or containers
- Approximate dimensions of the largest crate and/or container, length, width, height
- FOB point of shipment (for imported components)
- Point of shipment (for domestic components)

SCHEDULE C7 — PRELIMINARY WORK PROGRAM

The Bidder shall submit a time bar chart designated the Work Program clearly indicating the contractor's proposed method and sequence of performing the work.

The Work Program shall show the timing of all significant events related to the contract such as the design, material and equipment procurement, fabrication, shop testing, and delivery. The Bidder shall clearly indicate those sections of the work to be carried out by subcontractors.

CONTENT OF THE BID (cont)

CENTRIFUGAL AIR COMPRESSORS—CONTRACT No. B/417

SCHEDULE —TECHNICAL

SCHEDULE T1 — DESCRIPTION OF THE TECHNICAL PROPOSAL

A full description of the equipment offered shall accompany the Bidder's bid. This description shall include information in sufficient detail to enable the equipment offered to be fully evaluated in regard to layout, design, construction, performance, maintenance and availability of spare parts.

SCHEDULE T2 — DRAWINGS AND TECHNICAL DATA SUBMITTED WITH BID

The below listed documents are included in the Bidder's bid in accordance with the drawing and technical data submittal requirements form.

DOCUMENT NO.	TITLE OF DOCUMENT

SCHEDULE T3 — DEVIATIONS FROM THE TECHNICAL SPECIFICATION

The Bidder shall set out below a tabulated statement showing clearly, and in order of the relevant clauses, any deviations from the technical specification.

DOCUMENT NO.	PARAGRAPH NO.	COMMENTS

CONTENT OF THE BID (cont)

CENTRIFUGAL AIR COMPRESSORS—CONTRACT No. B/417

SCHEDULE T4 — PROPOSED ALTERNATIVE BID

The Bidder may provide an alternate bid that he considers to be of greater advantage in meeting the Purchaser's requirements for the intended service of the equipment with respect to price, ease of operation, maintenance and availability of spare parts. Such bid will only be considered when a conforming bid has been submitted and shall be clearly marked "Alternative Bid." The alternative bid shall be supported by a statement that clearly defines in detail the areas where the alternative offer differs from the bidding documents together with the cost reduction and other advantages to enable the Purchaser to make an appropriate assessment.

SCHEDULE T5 — COMPONENTS NOT SUBJECT TO THE EQUIPMENT WARRANTY OR DEFECTS LIABILITY

The Bidder shall list below all component parts that require replacement due to wear in normal operation and would not be included in the equipment warranty.
This list is subject to the Purchaser's approval prior to award of contract.

NAME OF COMPONENT	REFERENCE DRAWING

SCHEDULE T6 — MAINTENANCE TOOLS AND DEVICES

The Bidder shall list below maintenance tools and devices to be supplied as part of the contract, each of which has been included in the contract price. The Purchaser is not bound to purchase all or any of the tools or devices set out below.

NAME OF TOOL OR DEVICE	NUMBER TO BE SUPPLIED	CONTRACT VALUE $US

CONTENT OF THE BID (cont)

CENTRIFUGAL AIR COMPRESSORS—CONTRACT No. B/417

SCHEDULE T7 — ESSENTIAL SPARES

The Bidder shall list below the particulars, quantities, and values of the essential spares required for 9,000 hours of operation of each unit of the equipment to be supplied as part of the contract, the value of which is included in the contract price. The Purchaser exercises the right to add or delete from the items listed as essential spares at the prices shown.

ITEM NO.	QUANTITY PER ITEM	PARTICULARS OF ITEM	CONTRACT VALUE $US

SCHEDULE T8 — QUALITY CONTROL

The Bidder shall outline in his bid, the quality control system that will be followed for the fabrication of equipment components under this contract. The statement shall contain:

- Whether the contractor's and subcontractor's manufacturing facilities are ISO 9000 registered.
- A description of the contractor's in-house quality control systems (inspection and tests) as well as subcontract control.
- What contractor's quality control requirements are specified in his purchase orders on subcontractors.
- What surveillance is made at the subcontractor's premises and at what intervals.
- Documentation and quality records, including material supplier's inspection reports, test results, and other quality certificates and reports.
- Whether inspection and test equipment is systematically calibrated in accordance with national standards. Whether the equipment test facilities operated by the contractor are recognized by a national certifying authority.
- The contractor's inspection procedures for final preservation, packaging, and shipping, to ensure that specified requirements are met.

CONTENT OF THE BID (cont)

CENTRIFUGAL AIR COMPRESSORS—CONTRACT No. B/417

SCHEDULE T9 — CONTRACTOR'S CAPACITY TO PERFORM

The Bidder shall provide the following information:

(a)   Contractor's company structure together with an interdepartmental organization chart.

(b)   Résumés of key personnel assigned to this contract, which are to include:
  - Present position and responsibilities
  - Previous positions and employment together with responsibilities
  - Professional, trade, and other qualifications

(c)   Copy of the latest audited financial statement of the contractor's company including the annual report.

(d)   A list of comparable work completed by the contractor or in progress.

Note: (a), (b), and (c) in this schedule T9 are not normally required from well known specialist equipment suppliers. However, this information would be needed for large procurements.

SCHEDULE T10 — PROPOSED SUBCONTRACTORS

Should the Bidder propose to sublet any portion or portions of the work, he or she shall list the names and addresses of all proposed subcontractors providing any part of the work having a value in excess of $US 20,000.

Subcontracting will not be permitted without the written approval of the Purchaser.

ITEM TO BE SUBCONTRACTED	SUBCONTRACTOR	LOCATION

CONTENT OF THE BID (cont)

CENTRIFUGAL AIR COMPRESSORS—CONTRACT No. B/417

SCHEDULE T11 — RECOMMENDED SPARE PARTS

The Bidder shall itemize any spare parts, other than spare parts specifically included under the contract, that are recommended for the Purchaser to acquire for 48 months operation of the equipment. These spares shall be fully itemized. Prices shall be quoted in $US currency, including packaging, and delivery into the Purchaser's warehouse.

ITEM DESCRIPTION	PART NO.	NUMBER RECOMMENDED	COUNTRY OF ORIGIN	UNIT PRICE	NET KG	DELIVERY TIME

15.1.4   Bid Form

BID FORM

To: (Name and Address of Purchaser ....
hereafter called "Purchaser")

....

....

....

1. The undersigned, having examined the proposed Bidding Documents titled:

(Contract Number)   ....

(Name of Equipment)   ....

hereby proposes and agrees to furnish all labor, materials, and equipment to perform the complete Work as required by the said proposed Bidding Documents including Supplemental Notices Nos. .... to ....
received, for the
stipulated sum of ....
....
(in words)

($..............................)

in accordance with the Price Schedules attached hereto and made part of this Bid.

2. The undersigned acknowledges that he has complied with the Instructions to
Bidders issued for this Work.

3. The undersigned agrees to execute the Work in accordance with the Contract
Program provided in the Contract Documents.

4. The undersigned agrees to abide by this Bid for a period of .... (number) days
from the date fixed for bid closing pursuant to the Instructions to Bidders for
Contract No.................., and it shall remain binding and may be accepted at any
time prior to the expiration of that period.

5. Until a Contract Agreement is prepared and executed, this Bid together with the
Purchaser's written Notification of Award of Contract, shall form a binding contract
between the Purchaser and the undersigned.

Page 1 of 2

BID FORM (cont)

6. The undersigned understands that the Purchaser is not bound to accept the lowest priced bid that the Purchaser receives.

Bidder:...

Sign "under hand"   Address: ....

or   ....

"under seal"   Signature: ....

Title/Position: ....

Dated this .... day of ....19 ....

Signature of Witness: ....

Title/Position: ....

15.1.5   Contract Agreement Form

Note: This document has important legal consequences. Consultation with an attorney is encouraged with respect to its wording and completion of content.

The Contract Agreement Form is attached to the bidding documents. After award of contract, it is to be executed by the two parties.

CONTRACT AGREEMENT

Contract Number: ....

This Contract Agreement is made .... day of ....19....... and entered into between Redman Chemicals Corporation, a corporation incorporated under pursuant to the laws of .... in .... with principal
(state)   (country)

offices in .... (hereafter called "Purchaser") and .... (hereafter called "Contractor")

Contract Agreement Scope: Contractor shall perform all of the Work described in and in strict accordance with this Contract Agreement and the contract documents which are listed below and by this reference incorporated as part of this Contract Agreement.

1. Notice of Award, or
   Letter of Acceptance of Bid   marked Exhibit "A"
2. Technical Requirements Form   marked Exhibit "B"
   (including all listed attachments)
3. Supplementary Contract Conditions   marked Exhibit "C"
4. General Contract Conditions   marked Exhibit "D"
5. The Contractor's Bid   marked Exhibit "E"

In case of conflict between any of the contract documents accompanying this Contract Agreement, the order of precedence shall apply in the same document order as noted above from 1. to 5.

Renumeration: In consideration of performance of the Work the Purchaser will make payments to the Contractor in accordance with the provisions of the contract documents.

This Contract Agreement shall take effect according to its intent notwithstanding any prior agreement in conflict or at variance with it or any correspondence or other documents relating to the subject matter of this Contract Agreement which may have passed between parties to this Contract Agreement prior to its execution and which are not included in the contract documents.

Page 1 of 2

CONTRACT AGREEMENT (cont)

Address:

Contractor:

Purchaser:
Redman ChemicalsCorporation
Montcliff Avenue
Reedtown CA 00012

IN WITNESS WHEREOF The parties have entered into this Contract Agreement in accordance with their respective laws and statutes or constitutions on the date first above written.

(Contractor)

(Purchaser)
Redman Chemicals Corporation

By...

By...

Title/Position: ....

Title/Position: ....

15.2   TECHNICAL DOCUMENTS FOR BIDDING

15.2.1   Technical Requirements Form

REDMAN CHEMICALS CORPORATION

TECHNICAL REQUIREMENTS FORM

TITLE Centrifugal Air Compressors   **CONTRACT No** B/417

PROJECT No 7780

DELIVER TO   Redman Chemicals Corporation   **REVISION No**   0

Altona Plant

Montcliff Avenue   **SHEET No** 1   **OF** 1

Reedtown CA 00012

ATTACHMENTS

	THE FOLLOWING DOCUMENTS FORM PART OF THE TECHNICAL SPECIFICATION		REV
1	Drawing and Technical Data Submittal Requirements Form		0
2	Centrifugal Compressor Specification	SP – 1503	0
3	Standard General Data and Requirements Specification	SSP – 0201	1
4	* Standard Electric Motor Specification	SSP – 3302	3
5	* Standard Noise Specification	SSP – 5101	2
6	* Standard Painting Specification	SSP – 5603	5
7	Centrifugal Air Compressor Data Sheet	DS – 1503 – 1	0
8	Induction Motor Data Sheet	DS – 3302 – 1	0
9			
10			

REV	ITEM NO	QTY & UNIT	DESCRIPTION	ACCOUNT NO
	1	2	Centrifugal air compressors with electric	
			motor drivers complete with all accessories	
			as specified	
			Tag Equipment: 1501 and 1502	7780/4
			* Assumed to be in Bidder's possession	

0	Mar 7,'92	Issued for bids	T. White	KTK	
REV	DATE	ISSUE DESCRIPTION	BY	APPROVALS	

15.2.2   Drawing and Technical Data Submittal Requirements Form

REDMAN CHEMICALS CORPORATION
DRAWING AND TECHNICAL DATA SUBMITTAL REQUIREMENTS

Contract No. B / 417   Sheet 1 of 1   Revision No. 0

PRINTED MATERIAL AND DRAWINGS TO SUPPLIED BY THE CONTRACTOR AS INDICATED

	TECHNICAL DATA FOR: Centrifugal Air Compressor Tag No: 1501 1502	PRINTS WITH BID	REVIEW CODE (1)	No. of DWGS TRANS-P'ANCY	PRINTS	No. of PRINTED MATERIAL	TIME/DATE REQUIRED (2)
1	Drawing List & Drawing Schedule			1			2
2	General Arrangement Drawings	3		1			4
3	Dimensional Outline & Details	3		1			4
4	Equipment Lifting Mass (kg)	3		1			4
5	Foundation Plan / Anchor Bolts			1			4
6	Load Diagrams – Foundation and Piping Forces and Moments			1			4
7	Lube Oil, Seal Oil, & Cooling Water Dwgs	3		1			6
8	Compressor Performance Curves	3					
9	Completed Data Sheets	3					
10	Mechanical Terminal Connection Data			1			6
11	Inlet Filter / Silencer Outline			1			6
12	Blow – off Silencer Outline			1			6
13	Electrical Schematic & Wiring Dwgs			1			6
14	Location / Size of Elect. Connections			1			6
15	Instrument & Control Panel Layout			1			6
16	List of Instruments & Controls	3		1			4
17	Vibration Probes & Schematic Dwgs			1			6
18	Installation Instructions		R1				4 ** 6 weeks B/S
19	Operating Instructions		R1				4 ** 6 weeks B/S
20	Maintenance Manuals and Complete Parts List		R1			4 **	6 weeks B/S
21	Certificates – Statutory Approval(s)		F			3	
	– Performance Test(s)		S			3	B/S
	– Shop Inspection Reports		S			3	B/S
22	Spare Parts with Delivery Times *	3					
	* Part of the Contract						

** First copy required 6 weeks B/S. Four final copies required

Legend : (1) Review Code F :   Review is required prior to fabrication
                      S :   Review is required prior to shipment
                      R1 :   One copy only to be submitted for first review
                      I :   Required for information only
          (2) In weeks after award of contract, or "B/S" before shipment

All data unless otherwise specified, are to be directed to:
                      Attention: Mr. K.T. Kenwood, Project Manager
                                    Redman Chemicals Corporation
                                    Montcliff Avenue
                                    Reedtown CA 00012   KTK.

15.2.3   Equipment Design and Performance Specification

REDMAN CHEMICALS CORPORATION
MONTCLIFF AVENUE
REEDTOWN   CA 00012

ALTONA PLANT EXPANSION DEVELOPMENT

PROJECT No   7780

ENGINEERING SPECIFICATION

No   SP – 1503

CENTRIFUGAL COMPRESSOR

CONTRACT No   B / 417

Sheet 1 of 11

0	Feb 15 '92	Issued for bids	T. White	KK
ISSUE	**DATE**	**ISSUE DESCRIPTION**	**BY**	**APPROVALS**

REDMAN	ENGINEERING SPECIFICATION		TWW	КПК
CHEMICALS	CENTRIFUGAL COMPRESSOR		Sheet 2 of 11	
CORPORATION	Specification No. SP-1503	Contract No. B/417	Revision No. 0	

1. SCOPE

This specification covers the general requirements for the design, construction, inspection, testing, performance, and warranty of two motor driven multistage centrifugal compressors for compression of oil-free air. This specification forms part of the Technical Requirements Form for the procurement of this equipment and shall be read in conjunction with all documents listed in the Attachments therein.

Throughout this specification, "Purchaser" refers to Redman Chemicals Corporation; "Contractor" refers to the party or parties submitting a bid or supplying the equipment specified herein; "Manufacturer" refers to a concern that manufactures a part or parts of the equipment or material. (Contractor and Manufacturer may be synonomous).

Should the Contractor's interpretation of this specification suggest a conflict between it, the data sheets, or any other information contained in the contract documents, the Purchaser on request from the Contractor, will provide clarification before the Contractor proceeds with any work.

The order of technical precedence shall be as follows:

1. Compressor and Motor Data Sheets DS-1503-1 and DS-3302-1
2. This Specification
3. General Data and Requirements Specification SSP-0201
4. Standard Specifications listed in the Technical Requirements Form

2. GENERAL REQUIREMENTS

The Contractor shall furnish each compressor unit and its accessories in conformance with the data sheets and as specified in the following paragraph sections to make a complete installation.

The compressor units will be located out-of-doors with only a canopy type roofing to be furnished by the Purchaser. The installation environment is classified as a nonhazardous, corrosion-free area but shall be protected against infiltration of moisture and dust and shall be suitable for operation under high winds, rain, and humidity prevailing at the site.

The compressors will operate in parallel for a continuous 12-month period with shutdown only for routine maintenance. The units shall be suitable for unattended operation with pushbutton start and stop control from both local and remote (central control room) stations. Each unit shall be suitable for controlled stepless operation to meet discharge flow rates varying from 60 to 105 percent of the rated capacity and pressure requirements while retaining optimum efficiency.

Each compressor shall be coupled to an induction motor driver and the complete compressor unit shall be mounted on a common baseplate supplied by the Contractor.

Intercoolers at each stage and aftercoolers shall be supplied to achieve high compressor efficiencies. It is mandatory that the shaft sealing system be furnished for absolutely oil-free compression.

REDMAN	ENGINEERING SPECIFICATION		TWW	KπK	
CHEMICALS	CENTRIFUGAL COMPRESSOR		Sheet 3 of 11		
CORPORATION	Specification No. SP-1503	Contract No. B/417	Revision No. 0		

A major requirement of this specification is the interchangeability of compressor parts and the ready availability of spare parts.

3. WORK BY CONTRACTOR

The Contractor shall be responsible for the following:

(a) Supply and delivery of two compressors and motor drivers complete, each with common rigid base
(b) Aftercooler to limit discharge temperature as specified
(c) Combined inlet filter-silencer for each compressor with inlet piping to each compressor
(d) Blow-off vent system with operator and silencer complete with interconnecting piping and pipe supports
(e) Accessories as specified
(f) Machine mounted control panel with instrumentation and control equipment as specified

4. PURCHASER'S RESPONSIBILITIES

The Purchaser will be responsible for the following:

(a) Unloading the equipment at site
(b) Installation of the equipment
(c) Canopy type roofing
(d) Design and construction of the foundations with anchor bolts
(e) Discharge air piping from Contractor supplied check valve for each compressor
(f) Supply of piping to one common inlet and one common outlet terminal point for cooling water for each compressor
(g) From a common terminal point, condensate piping from Contractor supplied moisture traps and drains
(h) Electrical wiring from motor terminal boxes to Purchaser supplied motor starters, and 120 V AC systems
(I) A 4 to 20 mA control signal to the Contractor's supplied compressor control system

REDMAN	ENGINEERING SPECIFICATION	TWW $\mathcal{KTK}$	
CHEMICALS	CENTRIFUGAL COMPRESSOR	Sheet 4 of 11	
CORPORATION	Specification No. SP-1503	Contract No. B/417	Revision No. 0

5.   APPLICABLE CODES, SPECIFICATIONS AND INDUSTRY STANDARDS

Standard and other publications, of the organizations listed below, form part of this specification in the indicated subject areas. The applicable issue shall be the one current at the date of submission of bid.

OSHA - Environmental and safety
NEMA - Electrical
NFPA - National Fire Protection Association
ASTM - Materials
AGMA - Gears
AWS   - Welding
ASME - Piping
ASME - Pressure vessels
CAGI   - Compressors

6.   COMPRESSOR CONSTRUCTION

6.1   Materials of Construction

The materials of construction shall be as specified on the Centrifugal Air Compressor Data Sheet DS-1503-1.

6.2   Casing

The complete casing design pressure shall cover the highest possible operating pressure which normally will fall at rated speed and surge. The allowable external forces and moments on piping connections shall be shown on the contractor's outline drawing.
    The compressor casing shall be hydrostatically tested at 150 percent of the casing design pressure.
    No repairs are permitted on cast iron casings.

6.3   Rotating Element

The design of the compressor shall be such that inspection of impeller and diffuser (if used) shall be possible without disturbing the intercoolers, the alignment of the driver or the main sections of the air inlet and discharge piping except elbows and expansion joints.
    Impeller material shall be made of wear and corrosion resistant stainless steel.
    All rotating parts shall be dynamically balanced and shall be overspeed tested at 110 percent of maximum operating speed. Shaft with impeller(s) and pinion assembled shall be dynamically balanced.
    Axial impeller thrusts shall be absorbed by tilting pad thrust bearings. Anti-thrust bearings shall be supplied to provide safety in case of compressor surging. Thrust bearings shall be sized for equal loads in both directions.

REDMAN CHEMICALS CORPORATION	ENGINEERING SPECIFICATION CENTRIFUGAL COMPRESSOR	TWW $\mathcal{KK}$
		Sheet 5 of 11
	Specification No. SP-1503	Contract No. B/417	Revision No. 0

6.4 Bearings

Journal bearings shall be tilting pad sleeve type, pressure lubricated. Antifriction type bearings are not acceptable.

The mechanical arrangement shall be such that inspection of bearings and probes shall be feasible without disturbing the intercooler or driver alignment.

6.5 Seals

The compressor seals shall prevent leakage of oil-free air by providing separate oil and air seals, preferably horizontal split. Seals shall be self-lubricating. A buffer air sealing system shall be furnished if necessary to maintain the compressor oil-free air during start-up operation.

Seals shall be removable without opening the compressor casing.

6.6 Coupling

The couplings between the compressor gearbox and motor driver shall be an all metal, flexible gear type, or metal flexible disk type, be suitably sized and have sufficient capability to transmit the maximum motor power with an appropriate safety factor.

The couplings shall permit easy removal of either driver or driven component during opening the compressor for maintenance and inspection purposes.

The couplings shall be dynamically balanced independently from the rotating assemblies to a tolerance suitable for the maximum continuous speed.

A removable, heavy duty type coupling guard shall be furnished. This guard shall be in accordance with the statutory requirements prevailing at the site.

6.7 Baseplate

A fabricated structural steel baseplate shall be furnished to mount the complete compressor unit including motor driver, lubrication system, inter/aftercoolers and instrumentation/control panel and interconnecting piping, wiring, and tubing. The construction of the baseplate shall be such that holes or lifting lugs be provided for lifting the compressor assembly without damaging the compressor or any related component.

6.8 Gearbox

The compressor gearbox shall conform to AGMA Standards and be preferably horizontally split for simple accessibility, inspection and maintenance of gears, bearings and seals.

Gears shall be hardened and designed in accordance with AGMA Standards. Lubrication shall be by a forced feed oil system. Gear housing shall be equipped with a vent and oil demister.

REDMAN	ENGINEERING SPECIFICATION	TWW *KTK*	
CHEMICALS	CENTRIFUGAL COMPRESSOR	Sheet 6 of 11	
CORPORATION	Specification No. SP-1503	Contract No. B/417	Revision No. 0

6.9   Lubrication System

The lubrication system shall form an integral part of the compressor unit and include but not be limited to the following:

(a)   Oil reservoir tank with fill connection and level indicator, vent connection with an oil mist arrestor and drain connection. The oil reservoir shall have a minimum of five minute retention time. The reservoir shall be furnished with pressure, temperature, and level indicators as well as level and temperature switches.

(b)   Lubrication oil heater complete with thermal controls

(c)   Main oil positive displacement pump to be shaft driven to supply the lubrication oil under pressure.

(d)   Auxiliary motor driven positive displacement lubrication oil pump for use when the compressor is running up from start and decelerating to a stop.

(e)   Twin full capacity coolers of the shell and tube type similar to the air coolers specified in Section 6.10. Temperature gauges shall be furnished on the oil inlet and outlet piping.

(f)   A full capacity oil filter unit of the replaceable cartridge type with quick acting three-way valves for unit change-over. The filter system shall have a pressure differential gauge/switch.

(g)   All piping shall be carbon steel. Screwed connections to the filters, coolers, pumps and other components shall have unions for easy removal. Piping shall be cleaned and pickled prior to assembly. After assembly, all piping systems shall again be thoroughly cleaned and flushed.

(h)   Instrumentation and accessories such as relief valves, pressure gauges and switches shall be supplied.

(I)   Motor bearing lubrication, if required.

6.10   Intercoolers and aftercoolers

The Contractor shall provide a single inlet and a single outlet cooling water connection to serve both the intercooler(s) and aftercooler for each compressor.

The design of the inter and after coolers shall be in accordance with the ASME Code for Pressure Vessels. The aftercooler shall be designed for 8 to 10 °C air discharge approach temperature with 26 °C inlet water temperature.

Slide-out intercoolers and aftercoolers shall be of the shell and tube type with horizontal straight through removable tube bundles. Tubes shall be copper-nickel or similar material. The tube bore diameter shall not be less than 15 mm.

The maximum and minimum permitted water velocity shall be 3 m/s and 1 m/s respectively.

Intercoolers and aftercoolers shall be readily accessible for tube cleaning, maintenance and removal of tube bundles without disturbing air, cooling water or drain piping.

The coolers shall be equipped with suitable demisting devices to prevent liquid carryover.

REDMAN	ENGINEERING SPECIFICATION	TWW	KTK
CHEMICALS	CENTRIFUGAL COMPRESSOR	Sheet 7 of 11	
CORPORATION	Specification No. SP-1503	Contract No. B/417	Revision No. 0

Separation traps shall be supplied to continuously drain condensate and shall have stainless steel or nonferrous mechanical drain parts to prevent corrosion and assure continuous proper operation of the moving parts. Drain valves shall be carbon steel with stainless steel or nonferrous trim. Drain lines shall be minimum 12 mm bore diameter.

7. ACCESSORIES

The following accessories shall be furnished with each compressor unit:

(a) Two-stage fixed-element inlet air filter, including weatherhood
(b) Flexible connections for compressor inlet and discharge
(c) Check valve at compressor discharge
(d) Cooling water control system
(e) Continuous draining system including condensate traps, drainers, and bypass valves, shop mounted and piped
(f) Vibration monitoring equipment
(g) Special tools and fixtures to assemble and disassemble the compressor

8. MOTOR DRIVER

The compressor motor with all accessories shall be in accordance with Motor Data Sheet DS-3302-1.
The compressor motor shall be designed for service on Purchaser's three phase, 4160 Volt, wye-connected, grounded system, and have a NEMA standard enclosure for outdoor service, with terminal box one size larger than standard.
The compressor motors shall be designed and tested in accordance with latest NEMA Standards, including variations in voltage and frequency. Their design shall be such as to withstand full voltage starting, with low starting torque and low starting current.
The compressor motor shall be rated to develop 110 percent of the maximum power necessary to meet all operating conditions. The losses shall be included in the compressor power requirements.
The power kW rating shall be based on operation at full load without exceeding the rated temperature rise specified in the induction motor data sheet.
The motor shall be mounted on the common compressor base and be prealigned with the compressor in the Contractor's shop.
Each compressor motor shall be equipped with the following accessories:

(a) Embedded temperature detectors, two per phase, brought out and connected to an external terminal box, then wired to the main control panel
(b) Bearing temperature detectors connected to the control panel
(c) Internal space heaters brought to above terminal box, suitable for connection to the Purchaser's 120 Volt, single phase, 60 Hz power supply

REDMAN	ENGINEERING SPECIFICATION	TWW *KK*
CHEMICALS	CENTRIFUGAL COMPRESSOR	Sheet 8 of 11
CORPORATION	Specification No. SP-1503 Contract No. B/417	Revision No. 0

(d)   Foregoing terminal box shall be separate from the motor terminal box, but shall be located adjacent thereto

The motors shall have suitable grounding pads and shall have one or both bearings insulated to prevent shaft currents. The motor enclosure shall have one plugged drain hole in each end bell.

9.   NOISE LIMITATION

The compressor system shall be guaranteed to operate with a sound level not exceeding a 85 dB(A) slow response for an eight hour exposure period including sound emitted from the blow-off system. Measurements shall be taken at points one meter from the equipment in accordance with the procedure set forth by CAGI-PNEUROP Test Code.

A noise suppression enclosure shall be supplied if the compressor system exceeds the above sound level limitation. The design of the enclosure shall be subject to the Purchaser's approval prior to fabrication.

The noise sound level from the vent blow-off system shall not exceed 85 dB(A) when measured one meter from the equipment.

10.   INSTRUMENTATION AND CONTROL EQUIPMENT

10.1   General

The Contractor shall supply for each compressor, all instrumentation and a microprocessor based control for safe, efficient and reliable operation of the unit. The instrumentation and control equipment shall be prewired, installed, and tested in the contractor's shop. All control valves, bypass valves, transducers, and similar equipment shall be mounted on the compressor unit.

The microprocessor control system shall be designed for unattended continuous operation of the compressor units. The surge control system shall avoid surge under all operating conditions and shall be complete with control valves and control actuators. The control shall prevent overload of the motor driver for all specified operating conditions.

Each high-speed bearing shall have two noncontacting type vibration probes, installed in the same plane, 90 degrees apart, wired into a common shutdown device in order to trip the unit on excess vibration. This trip switch will be wired to the Purchaser's motor starter; wiring and starter being supplied by the Purchaser.

Indicating instruments shall be clearly visible from normal operating positions. Permanent access shall be provided by the Contractor where operator adjustments have to be made to the equipment.

The Contractor shall supply all instrument wiring and tubing between machine-mounted instruments and the control panel.

REDMAN	ENGINEERING SPECIFICATION	TWW	*KTK*	
CHEMICALS	CENTRIFUGAL COMPRESSOR		Sheet 9 of 11	
CORPORATION	Specification No. SP-1503	Contract No. B/417	Revision No. 0	

10.2   Compressor Controls

Compressor controls shall be included but not be limited to the following:

1.   A selector switch to allow the compressor to operate in the following control modes:

(a)   Unload—Inlet valve closed, blow-off valve open
(b)   Modulate control—remain constant pressure for load variations by throttling the air inlet vane and venting to atmosphere
(c)   Automatic dual control—maintain constant pressure by throttling the air inlet at the minimum turndown point and if demand is less than compressor output, the compressor shall unload by closing the inlet and opening the blow-off valve. The compressor shall reload upon falling air pressure in the pipe system.

2.   Operating modes to be provided are:

(a)   Automatic loading on start-up
(b)   Maximum base load operation
(c)   Base load/unload pressure operation
(d)   Intermittent load operation

3.   The microprocessor control system shall display and provide alarms and shutdowns but not be limited to the following:

	Display	Alarm	Shutdown
Air discharge temperature at each stage high	X	X	X
Lube oil temperature high	X	X	X
Lube oil pressure low	X	X	X
Lube oil filter - pressure drop high	X	X	
Lube oil reservoir level - high/low			X
Main oil pump pressure high	X		
low	X	X	X
Shaft vibration monitor at each stage	X	X*	X*
Air discharge pressure at each stage		X	

*when excessive

REDMAN CHEMICALS CORPORATION	ENGINEERING SPECIFICATION CENTRIFUGAL COMPRESSOR		TWW	ĸπĸ
	Specification No. SP-1503	**Contract No. B/417**	**Sheet 10 of 11**	
			Revision No. 0	

	Display	Alarm	Shutdown
Bearing oil pressure high	X	X	
low	X	X	X
Surge			unload
Seal air	X		X
Loss of cooling water (flow switch)	X		X
Air filter - high pressure drop		X	
Auxiliary lube oil pump running	X		
Main motor current		X*	X*
Panel power on	X		
Permissive start	X		
Service hour meter	X		

*when excessive

4. Unload compressor at surge
5. Shutdown the auxiliary lubrication oil pump when main (shaft driven) pump attains the operating pressure at the bearings; and start-up if main lubrication oil pump pressure drops
6. Interlocks to prevent start-up of compressor on low lubrication oil pressure, or on low cooling water flow (monitored at cooler outlets)

The microprocessor panel shall have annunciator pushbuttons to acknowledge alarmed malfunctions. The panel shall also have connections to wire the alarm and shutdown switches to the Purchaser's remote central control room instrument panel.

10.3   Manual Adjustments and Settings at Microprocessor

The following facilities are considered as a minimum to be provided at the microprocessor for control adjustments:

(a) System air circulation pressure
(b) Control valve actuator including settings for proportional band, integral time and reload

11.   INSPECTION

Refer to General Data and Requirements Specification SSP - 0201

REDMAN CHEMICALS CORPORATION	ENGINEERING SPECIFICATION		TWW	ктк
	CENTRIFUGAL COMPRESSOR		Sheet 11 of 11	
	Specification No. SP-1503	Contract No. B/417	Revision No. 0	

12. TESTS

Each compressor unit complete with its own motor driver and mounted on its own baseplate shall be given a full operating shop test to demonstrate mechanical integrity, air temperatures, air flow capacity with corresponding discharge pressure, and power consumption.

The minimum test running period shall be two hours and test measurements shall be made in accordance with ASME Power Test Code PTC-10. During the test, the surge point and stone wall at design operating speed shall be demonstrated. A test curve consisting of a minimum of four points shall be submitted to the Purchaser. Noise level measurements, and vibration levels shall be recorded during shop tests.

The Bidder shall guarantee the maximum power consumption.

13. NAMEPLATES AND TAGGING

Refer to General Data and Requirements Specification SSP-0201

14. PAINTING AND PROTECTIVE COATINGS

Refer to General Data and Requirements Specification SSP-0201

15. ASSEMBLY, PACKING AND SHIPPING

Refer to General Data and Requirements Specification SSP-0201

16. WARRANTIES

16.1 Equipment Warranty

The Contractor shall warrant the compressor unit for a period of two years from the date of acceptance by the Purchaser that all equipment furnished by the Contractor shall be free from defects in design, material, and workmanship.

Should servicing or replacement of parts be required during the warranty period due to defective design, material, or workmanship, new replacement parts shall be furnished and installed promptly by the Contractor at no additional cost to the Purchaser.

16.2 Performance Warranty

The compressor shall be performance warranted for pressure, capacity and power requirements. The compressor characteristics shall be such that the two units can run in parallel between 60 and 105 percent of rated load without hunting, surging, or other undesirable operation.

15.2.4   Equipment Data Sheets

#		
1	Compressor Manufacturer	Model
2	Electric Motor Furnished by : Contractor	Motor Data Sheet No DS – 3302 – 1
3	SITE CONDITIONS	
4	Elevation 150 m Ambient Temperature Max / Min / Design: 40 / – 5 / 30	° C
5	Environment : Windy and Dusty	Intallation : Indoor / Outdoor
6	Inlet Cooling Water Pressure 480 kPag	Inlet Cooling Water Temperature 26 ° C
7	GUARANTEED COMPRESSOR PERFORMANCE	
8	INLET CONDITIONS	DISCHARGE CONDITIONS
9		Pressure 860 kPag
10	Inlet Pressure	Temperature ° C
11	Barometric Pressure 99.2 kPaA	
12	Temperature 30 ° C	
13	Relative Humidity 60 %	
14	Inlet Volume 250 m ³ / min	
15	Cooling Water Temp 26 ° C	Max Cooling Water Temp Rise ° C
16	Power Required (all losses included)	kW
17	Input Speed rpm	
18	Driver Power Required kW	
19	Estimated Surge m ³ s	
20	Efficiency %	
21	Total Cooling Water Flow L / min	
22	Performance Curve No from 10 % to 110 % of Rated Compressor kW	
23	CONTROL	
24	Source : Purchaser Type : Electronic Range : 4 to 20 mA	
25	Control Operating Range 60 % to 105 %	
26	COMPRESSOR AIR SIDE DESIGN	
27	Design Pressure kPag	Max. Operating Temperature ° C
28	Hydrostatic Pressure kPag	Minimum Wall Thickness mm
29	MASS AND DIMENSIONS	
30	Complete Compressor Package with Driver	kg
31	Mass of Largest Part for Maintenance Lift	kg
32	Dimensions: Length mm / Width mm / Height mm	
33	APPLICABLE REFERENCE SPECIFICATIONS	
34	Compressor SP – 1503	Noise Refer to Compressor Specfication
35	Electric Motor SSP – 3302	Painting SSP – 5603
36	Notes:	
37		
38		

R						
E						
V	0	Mar 2, '92	Issued for bids – Bidder to complete these 3 sheets	JCT	KJK	
	No	DATE	ISSUE DESCRIPTION	BY	APPROVALS	

REDMAN CHEMICALS CORPORATION	**CENTRIFUGAL AIR COMPRESSOR** **EQUIPMENT DATA SHEET**	DOCUMENT No
		DS-1503-1
	TAG No 1501, 1502	SHEET 1 of 3
	CONTRACT No B/417	REVISION No 0

1	CONSTRUCTION				
2	MECHANICAL DESIGN				
3		Stage 1	Stage 2	Stage 3	Stage 4
4	Max Impeller Diameter mm				
5	Max Tip Speed m / s				
6	Critical Speeds rpm				
7	Connections:	Size	Rating	Facing	Orientation
8	Inlet				
9	Outlet				
10	Blow – off				
11	C. W. Inlet				
12	C. W. Outlet				
13	AIR INTERCOOLERS AND AFTERCOOLER				
14		Stage 1	Stage 2	Stage 3	Aftercooler
15	Type				
16	Cooling Area Air / Water Side m ^{2}				
17	Tube Material				
18	Tube O D / I D mm				
19	Number of Tubes				
20	Water Flow L / s				
21	Water Velocity m / s				
22	MATERIALS		CONSTRUCTION		
23	Impellers		Impellers – Closed / Semi Open		
24	Casing		Radial / Backward Lean		
25	Scrolls		Casing – Cast / Fabricated		
26	Nozzles		Scrolls – Cast / Fabricated		
27	Diffusers		Diffusers – Vaned / Not Vaned		
28	Diaphragms		Diaphragms – Cast / Fabricated		
29	Gear Housing		Gear Housing – Cast / Fabricated		
31	Pinion(s)		Pinions – Fabricated / Forged		
32	Shaft		Bull Gear – Cast / Fabricated / Forged		
33	Bull Gear		Journal Bearings		
34	Journal Bearings		Pinion – Type		
35	Pinion		Bull Gear – Type		
36	Bull Gear		Thrust Bearings		
37	Thrust Bearings		Pinion – Type		
38	Pinion		Bull Gear – Type		
39	Bull Gear		Labyrinth		
40	Compressor Seals		Base Plate – Common / Compressor Only		
41					
42					
43					

REDMAN CHEMICALS CORPORATION	CENTRIFUGAL AIR COMPRESSOR EQUIPMENT DATA SHEET	DOCUMENT No
		DS-1503-1
	TAG No 1501, 1502	SHEET 2 of 3
	CONTRACT No B/417	REVISION No 0

#			
1	CONSTRUCTION		
2	LUBRICATION SYSTEM		
3	Inlet Oil Temp – Nom	° C	Oil Cooler – Number Supplied:	
4	– Max	° C	Number of Passes:	
5	– Min	° C	Water Flow / Velocity / m / s	
6	Reservoir Capacity	L	Number of Tubes / Length / mm	
7	Retention Time	min	Tube I D / O D / mm	
8	Auxiliary Pump	kW	Tube Material	
9	– Enclosure:		Tube Sheets / Baffles: /	
10	– Voltage / Phase / Hz:		Shell Dia / Length / Thickness mm	
11	Filter – Type:		Water Temperature Rise ° C	
12	– Micron:		Design Pressure: Shell kPag	
13	– Element:		Tubes kPag	
14	Lube Oil Heater –	kW	Lube Oil Piping Material	

#	ALLOWABLE PIPING FORCES AND MOMENTS		SHOP INSPECTION & TESTS		
15					
16		Forces N	Moments N • m	* to be certified	Req'd *	Witness
17	Inlet Axial			Shop Inspection	yes	no
18	Vertical			Hydrostatic	yes	no
19	Horiz 90 Deg			Mechanical Run	yes	no
20	Discharge Axial			Performance		
21	Vertical			Test with Driver		yes
22	Horiz 90 Deg			Vibration Probes	yes	no
23	Bi – Pass Axial			Bearings and		
24	Vertical			Seals after Test	yes	no
25	Horiz 90 Deg			Noise Level	yes	no
26					

#	ASSESSORIES		
27			
28	Type	Make	Model
29	Coupling		
31	Vibration Probes		
32	Axial Probes		
33	Vibration Monitor		
34	Vibration Readout		
35	Inlet Control Valve		
36	Unloading Valve		
37	Air Filter / Silencer		
38	Blow – off Silencer		
39	Discharge Check Valve		
40			
41	Notes:		
42			
43			

REDMAN CHEMICALS CORPORATION	**CENTRIFUGAL AIR COMPRESSOR** **EQUIPMENT DATA SHEET**	DOCUMENT No DS–1503–1
	TAG No 1501, 1502	SHEET 3 of 3
	CONTRACT No B/417	REVISION No 0

1	SITE CONDITIONS		
2	Location Environment: Inland / ~~Seaboard~~ / Dusty and Windy / ~~Corrosive~~		
3	Ambient Temperature: Max / Min / Design 40 / – 5 / 40 ° C	Altitude 150 m	
4	BASIC DATA AND REQUIREMENTS		
5	Purchaser's Motor Specification No	SSP – 3302	
6	Manufacturer		
7	Type		
8	Frame Designation		
9	Rating Output (kW)		
10	Rated Voltage	4160	
11	Phases / Frequency	3 / 60	
12	Service Factor		
13	Type of Construction		
14	Method of Cooling		
15	Protection		
17	Rated Current (amps) FL		
18	Rated Speed (rpm) Syn / FL		
19	Rotation – Viewed from drive shaft end		
20	Efficiency: 100%, 75%, 50% FL		
21	Power Factor: 100%, 75%, 50% FL		
22	Locked Rotor Current		
23	Rated Torque (Nm)		
24	Torque: Starting / Pull Out (% FLT)		
25	Inertia of Rotor		
26	Design Standard / Design Letter		
27	Enclosure		
28	Insulation Class		
29	Time Rating / Temp Rise ° C		
30	Winding Connection		
31	Bearings – Sleeve / Ball		
32	Type of Lubrication		
33	Slide Rails / Flange / Size		
34	Heaters: No. / Voltage / Rating		
35	Temperature Detect. Type / No / Temp		
36	Allowable Number of Starts / Hour		
37	Motor Mass (kg)		
38	Mounting Drawing No		
39	Termination Drawing No		

Right column illustration:

C
B   D
A   ◎   E

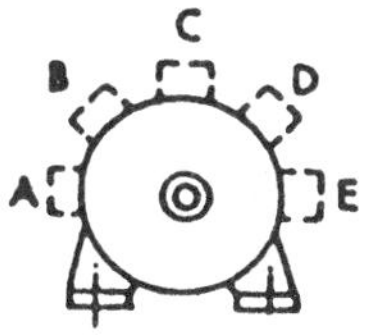

View on End Drive

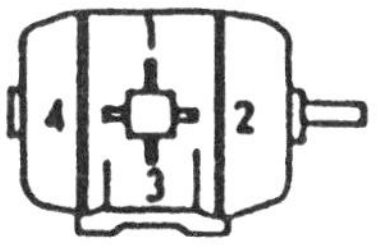

Side View

Location of Terminal Box and
Entry: A 3

40	Noise – Sound Pressure Level at 1 m @ Full Load dB(A)			
41	Motor Performance Curve No:	General Arrangement Dwg No:		
42	Witnessed Inspection and Testing: With Driven Equipment Yes / ~~No~~			
43	Furnished Options: Refer to Compressor Specification SP – 1503			

R E V	0	Mar 1, '92	Issued for bids – Bidder to complete this sheet	MBT	KK	
	No	Date	ISSUE DESCRIPTION	By	App'd	Date

REDMAN CHEMICALS CORPORATION	HORIZONTAL INDUCTION MOTOR DATA SHEET	DOCUMENT No DS-3302-1
	DRIVEN EQUIPMENT: Air Compressors 1501, 1502	SHEET 1 of 1
	CONTRACT No B/417	REVISION No 0

15.2.5   General Data and Requirement Specification

REDMAN CHEMICALS CORPORATION
MONTCLIFF AVENUE
REEDTOWN CA 00012

STANDARD
ENGINEERING SPECIFICATION

No   SSP – 0201

GENERAL DATA AND REQUIREMENTS

Sheet 1   of   10

ISSUE	DATE	ISSUE DESCRIPTION	BY	APPROVALS	
1	Feb 11 '92	Issued for bids	TWW	*KTK*	
0	Feb 5 '92	Revised as noted	JTC	KTK	
A	Jan 21 '92	Issued for in-house comment	TWW	KTK	

REDMAN	**ENGINEERING SPECIFICATION**		TWW	*KTK*	
CHEMICALS	**GENERAL DATA AND REQUIREMENTS**		**Sheet 2 of 10**		
CORPORATION	**Specification No. SSP-0201**	**Standard**	**Revision No. 1**		

1. SCOPE

This specification covers the general data and requirements applicable to equipment to be furnished by the Contractor in accordance with the Purchaser's contract documents.

Throughout this specification, "Purchaser" refers to Redman Chemicals Corporation; "Contractor" refers to the party or parties submitting a bid or supplying the equipment specified herein; "Manufacturer" refers to a concern that manufacturers a part or parts of the equipment or material. (Contractor and Manufacturer may be synonomous)

This specification shall be read in conjunction with the Attachments listed in the Technical Requirements Form. Should conflict arise between the provisions of this specification and other documents listed in the Technical Requirements Form, the Purchaser will provide written clarification upon request.

Unless specified otherwise, all equipment furnished by the Contractor shall be the manufacturer's standard product to permit early delivery, maximum service reliability and availability of spare parts. Consideration should be given in the selection of equipment, where economically practical, to provide maximum interchangeability of parts for units being supplied.

Contractor documents including nameplates and instructions shall have wording in the English language. Calculations, units for pressure, temperature, flows, drawing dimensions and the like shall be to International System of Units (SI).

2. GENERAL DESIGN DATA

2.1　Site conditions

The site is exposed to wind and dust. The equipment will be located outdoors and shall be designed for, and be protected against, the above prevailing conditions and the following:

Altitude	-	150 m
Ambient temperature, max./min.	-	40 °C / minus 5 °C
design	-	refer to equipment data sheet
Relative humidity, normal	-	60 %
Design wind force or design factor	-	Not applicable
Seismic zone factor	-	Not applicable

2.2　Power supply - voltage/phase/frequency

Motors	-	4160 V / 3 P / 60 Hz
	-	440 V / 3 P / 60 Hz
Other	-	120 V / 1 P / 60 Hz

REDMAN	ENGINEERING SPECIFICATION		TWW	ллл
CHEMICALS	GENERAL DATA AND REQUIREMENTS		Sheet 3 of 10	
CORPORATION	Specification No. SSP-0201	Standard	Revision No. 1	

Instrumentation and controls   -   manufacturer's standard

2.3   Compressed air pressure

Instrument (dry, oil free)   -   650 kPag
Service or utility   -   680 kPag

2.4   Water

Process (non drinking), temp./press.   -   26 °C / 480 kPag
Cooling (recycled), temp./press.   -   Not applicable
Potable, temp./press.   -   Not applicable

2.5   Noise level limitation   -   refer to equipment specification

3.   NATIONAL CODES, STANDARDS, AND STATUTORY REGULATIONS

The equipment supplied by the Contractor shall comply with all applicable ordinances, codes, regulations and laws pertaining to the Purchaser's site. Unless instructed otherwise by the Purchaser, the Contractor shall be responsible, and at his cost, for obtaining all necessary statutory approvals for the design and testing of equipment being furnished. Three (3) copies of each approval shall be forwarded to the Purchaser for record purposes. On the completion of the contract, the Contractor shall forward to the Purchaser, one (1) set of all statutory approvals, code and material test certificates received, properly bound between hard covers.

4.   MATERIALS AND WORKMANSHIP

All work shall be performed by skilled mechanics in the various crafts and shall be executed in a proper and workmanlike manner in accordance with recognized good practice.

Materials used in the fabrication of the equipment furnished by the Contractor shall be new and suitable for the intended services and be subject to the approval of the Purchaser. Equipment and materials when not specified in detail shall be in accordance with the best standard practice and as recommended by the supplier for the service.

5.   EXPEDITING, INSPECTING, AND TESTING

All equipment supplied by the Contractor shall be subject to expediting, inspecting, and testing during the duration of the contract. Each Contractor's suborder shall be endorsed: "subject to inspection and expediting by the Purchaser or his accredited representative."

REDMAN	ENGINEERING SPECIFICATION		TWW	ктк
CHEMICALS	GENERAL DATA AND REQUIREMENTS		Sheet 4 of 10	
CORPORATION	Specification No. SSP-0201	Standard	Revision No. 1	

The Purchaser reserves the right to inspect each item of equipment being fabricated under the contract and the right to review the Contractor's quality control procedures. Witnessing of quality control procedures and performance tests are only required when the equipment design and performance specification or the data sheet specifically states this requirement.

The Purchaser or Purchaser's representative shall have free access at reasonable times to the Contractor's works for the purpose of inspection of the work. The Contractor shall notify the Purchaser fourteen (14) calendar days in advance of the date that the equipment will be ready for final inspection, and/or performance test so that the Purchaser or Purchaser's representative can be present.

For purposes of expediting and inspections, the Purchaser requires one unpriced copy of each purchase order or subcontract issued by the Contractor and by each of his subcontractors.

Material and equipment that has been inspected by the Purchaser's inspector shall in no way relieve the Contractor of the responsibility to meet the requirements of the specification and all other provisions contained in the Purchaser's contract documents.

6. DRAWINGS AND TECHNICAL DATA SUBMITTAL REQUIREMENTS

The Contractor shall submit drawings and other data to the Purchaser for review as detailed and at the times indicated on the Drawing and Technical Data Submittal Requirements Form. This information will provide the Purchaser with a complete understanding of the equipment being supplied for the purpose of engineering the total plant such as preparing foundation drawings and equipment interfaces. Particular attention is drawn to the Contractor's notice that any fabrication commenced prior to the Purchaser's review shall be at the Contractor's risk.

All drawings shall form part of the contract after the Purchaser's review. The sequence of submission shall be such that all information is available for checking each drawing when it is received.

All drawings shall be presented on transparency film of 89 micrometers thickness, suitable for print production or microfilming. The title block shall contain the name of the Purchaser, contract name and number, and the Purchaser's equipment tag number together with the drawing number from a block of numbers issued by the Purchaser. The Contractor's own drawing number may be shown elsewhere on the drawing except in the title block.

Review of the Contractor's drawings by the Purchaser shall not relieve the Contractor of any part of his obligation to meet all requirements noted in the contract documents or the responsibility for correctness of such drawings. The Contractor shall make all necessary changes in design that are required to make the work conform to the provisions of the contract without additional cost to the Purchaser. All final submitted drawings shall be brought to an as-built status and be endorsed with the wording "As Built" in the revision block.

REDMAN	ENGINEERING SPECIFICATION		TWW	*KTK*	
CHEMICALS	GENERAL DATA AND REQUIREMENTS		Sheet 5 of 10		
CORPORATION	Specification No. SSP-0201	Standard	Revision No. 1		

7.   OPERATING AND MAINTENANCE MANUALS

Operating and maintenance manuals shall be supplied by the Contractor to provide all information deemed necessary by the Purchaser for his operating and maintenance personnel associated with the plant or equipment throughout its life. The number of copies to be submitted is
shown in the Drawing and Technical Data Requirements Form. The documentation shall contain
instructions, drawings (reduced as necessary, but clearly legible), parts lists, catalogs, and any
other instructions that may be needed or be useful in the installation, maintenance, dismantling,
assembling and for identification when ordering replacement parts. The manuals shall be assembled and bound between hard covers to 216 × 279 mm paper size (or A4, *state which*).
Where the Contractor's and/or subcontractor's standard brochures are included, they shall be
clearly marked to indicate the part that has actually been furnished by the subcontractor.

Each set of manuals shall be divided in separate volumes for:

- Installation instructions
- Operating instructions
- Maintenance manuals with list of spare parts

The manuals shall include, but not necessarily be limited to, the following:

(a)   All general information relating to major component overall dimensions, mass, lifting details and the like as required for assembly, installation and overhaul.

(b)   General description of the equipment including:

Design and material limits such as loadings, temperatures, out of balance limits and running clearance;

Grade of lubricants and frequency of lubrication;

A listing and identification of special tools and spare parts;

Serial numbers of all equipment.

(c)   Detailed description for placing equipment into and out of service including:

Commissioning procedures with settings and adjustments;

Precautionary checks before and during running; alarm and trip settings of relays;

Emergency procedures following an unscheduled stoppage of equipment;

Fault finding diagnostic charts for equipment failure during operating sequences.

REDMAN	ENGINEERING SPECIFICATION	TWW *KK*
CHEMICALS	GENERAL DATA AND REQUIREMENTS	Sheet 6 of 10
CORPORATION	Specification No. SSP-0201	Standard	Revision No. 1

(d)   General description detailing the operation of the equipment.

(e)   Details of the electrical circuitry accompanied by schematic diagrams; list of electrical interlocks and statement of their function.

(f)   A complete and detailed annual program for preventive maintenance covering all equipment supplied under the contract with due regard to hours of continuous equipment operation and the like.

(g)   Step by step instructions in sufficient detail to enable overhaul and replacement of all parts to be carried out. These instructions are to be grouped separately in sections to cover mechanical, electrical, and instrumentation maintenance.

8.   NAMEPLATES AND MARKINGS

Nameplates, notices and the like shall be permanently attached to the equipment to bring the attention to the following:

(a)   Notices required by statutory regulations such as maximum safe working load for hoists, lifts, and cranes

(b)   Warning notices such as "danger"

(c)   Equipment nameplates with manufacturer's model and data

(d)   Direction arrows for flow, rotation, "lift here" notices, and mass in kg

Suitable nameplates shall also be attached to valves and gauges furnished with the equipment.
    Nameplates and notices shall be of appropriate size and the information contained thereon be easily read from all operating positions. They shall be affixed to the equipment so that they cannot be removed or obliterated through wear and tear. Stampings and engravings shall be of sufficient depth so that the letters, numbers and wording shall not be worn away by the prevailing environmental conditions. Nameplate material shall be stainless steel or approved equal.
    Wherever possible, direction arrows shall be integrally cast into the body of the main casting.
    Equipment nameplates shall have all necessary information to facilitate identification and installation including:

- Name of manufacturer
- Purchaser's contract number
- Equipment tag number
- Manufacturer's serial number

REDMAN CHEMICALS CORPORATION	ENGINEERING SPECIFICATION		TWW	*KK*
	GENERAL DATA AND REQUIREMENTS		Sheet 7 of 10	
	Specification No. SSP-0201	Standard	Revision No. 1	

- Design and performance information as applicable
- Date of manufacture

9.   SPECIAL TOOLS AND DEVICES

The Contractor shall furnish one set of necessary special tools and devices required for the installation, operation, and maintenance of the plant or equipment supplied by him. These shall be new, be of first-class quality, crated separately from other equipment and materials, and be clearly marked as to their intended use.

The list of special tools and devices supplied shall be noted in the maintenance manuals.

10.   PAINTING AND PROTECTIVE COATINGS

Painting and protective coatings shall be in accordance with the applicable paint system specified in the Purchaser's standard painting specification. Where no specific instructions have been noted, the Contractor shall be responsible for applications detailed in the following paragraphs.

All equipment normally factory painted shall be smooth, free from fins, burrs, and other surface defects and cleaned to remove mill scale, rust, and other foreign substances before painting. Such surfaces shall be prime coated within four hours of cleaning and finished with a top coat of the manufacturer's standard color paint.

Machined and internal surfaces of equipment surfaces subject to corrosion during shipment and storage prior to installation, shall be coated with a corrosion inhibitor and be capable of providing protection during transportation and storage at the job site. The coating shall be readily removable with a commercial solvent.

Unless otherwise specified in the Purchaser's painting specification, procedures, paints used, and finish thickness shall be in conformance with the paint manufacturer's recommendation but be subject to the Purchaser's approval.

11.   PACKING AND SHIPPING

The Contractor shall furnish and install all bracing, shoring, and crating required. Unless otherwise specified, the equipment shall be strongly packed in accordance with the best commercial practice for shipment and transportation. Any damage received in transit due to defective or insufficient packing or defective methods of packing shall be made good by the Contractor at his expense and within a reasonable time when called upon to do so.

Each shipping piece and container shall be durably marked to show:

- Name and address of Purchaser
- Purchaser's contract number
- Name of equipment and equipment tag number(s)

REDMAN	ENGINEERING SPECIFICATION	TWW	KTK
CHEMICALS	GENERAL DATA AND REQUIREMENTS		Sheet 8 of 10
CORPORATION	Specification No. SSP-0201	Standard	Revision No. 1

All openings shall be suitably covered or plugged to prevent entrance of vapors or dust during shipment and storage at the job site. Flanged connections for piping shall be protected against corrosion by the application of a suitable, easily removable, rust preventative and shall have wood or steel cover plates bolted in place to protect the faces during shipment. Those connections equipped with permanent blind flanges or covers shall be bolted with the specified gasket in place ready for service.

Threaded connections shall be protected with brightly colored plastic thread protectors to be fitted before any required painting.

If the Contractor is responsible for the shipment of the equipment to the job site, he shall obtain the necessary approvals from the traffic authorities for wide-load clearance, high-load clearance and maximum mass load, for overland transportation to the job site. The maximum mass of any individual shipping item shall not exceed 15 tonnes.

The Contractor shall take particular note of quarantine regulations that apply at ports of destination. The use of straw for packing may not be permitted and all timber used must be free of insect infestation. The Contractor shall bear all costs such as fumigation and the replacement of packing as demanded by quarantine officials, even where the cost of shipping is not part of the Contract.

12.   EQUIPMENT WARRANTY

The Contractor shall warrant that the equipment supplied shall be entirely suitable for the intended service and is free from faults in design, and is of sufficient size and capacity to fulfill satisfactorily the operating conditions specified. The warranty period is specified in the Equipment Design and Performance Specification.

Should any item supplied by the Contractor be found defective after formal taking-over of the equipment, whether in design, workmanship or material, and providing the equipment has been operated in accordance with generally approved practice and in accordance with the conditions specified, the Contractor shall at his expense, correct all such defects without delay. If the Contractor is obliged to correct such defects, the warranty for the repaired or replaced part shall extend for the same period as noted above.

If the Contractor is unable or unwilling to repair or replace such defective part or parts within a reasonable time after having received written notice from the Purchaser, or if any emergency exists rendering it impossible or impractical for the Purchaser to notify the Contractor to repair or replace the defective part or parts, in all such cases the Contractor shall reimburse the Purchaser for the reasonable cost of replacing or repairing such defective part or parts.

13.   SPARE PARTS

The Contractor shall furnish spare parts and/or assemblies recommended to be carried in stock for efficient maintenance of the equipment being supplied. The quantity of spares shall be based on one years operation. Spares and assemblies supplied shall be packed and deliv-

REDMAN	ENGINEERING SPECIFICATION		TWW	КТК	
CHEMICALS	GENERAL DATA AND REQUIREMENTS		Sheet 9 of 10		
CORPORATION	Specification No. SSP-0201	Standard	Revision No. 1		

ered separately to the Purchaser, and be identified using the part numbers listed in the maintenance manuals.

Packages shall be clearly labeled with the following information:

- Purchaser's contract number
- Name of equipment supplied
- Name of spare part(s)
- Name of spare part(s) supplier(s)
- Supplier's spare part(s) number
- Number of items

The Contractor shall supply "construction spares" to cover normal predictable losses, wastage and damage during installation, testing and commissioning.

All spare parts supplied shall be listed in the maintenance manuals.

The following provisions are considered to be commercial rather than technical. They are included in this specification so as not to be omitted from the contract documents. Do not duplicate with other documents.

14.   WORK PROGRAM

Within two (2) weeks after being notified of award of contract, the Contractor shall submit to the Purchaser a completed bar chart detailing the scheduled major design, fabrication, and delivery activities.

Each item or activity on the bar chart shall have directly below the scheduled program, a sufficient horizontal space for monitoring "actual completion" status of each item and activity. The "actual completion" percentage status shall be noted at each monthly cut-off date for comparison between the "scheduled" progress of the item or activity and actual progress achieved.

The bar chart shall be monitored for actual progress against scheduled progress throughout the duration of the contract. Scheduled milestone dates shall not be altered or revised by the Contractor without the Purchaser's approval. If the milestone dates are changed, and if in the opinion of the Purchaser, delays have occurred that will affect contract completion, the Purchaser may request the Contractor to submit a revised program to replace the previous one issued.

Updated copies of the bar charts shall be appended to the progress reports (refer to Section 15 of this specification).

REDMAN	ENGINEERING SPECIFICATION		TWW	*KJK*
CHEMICALS	GENERAL DATA AND REQUIREMENTS		Sheet 10 of 10	
CORPORATION	Specification No. SSP-0201	Standard	Revision No. 1	

15. PROGRESS CONTROL AND REPORTING

The Contractor shall submit a detailed monthly progress report by the second Monday of each month giving details of all contract progress items. Each report shall be given under cover of a letter giving a summary of all aspects of the work including those carried out by subcontractors to the Contractor to update the Purchaser of the Contractor's overall progress in fulfilling his contractual obligations.

The format and detail of such progress report shall cover the following aspects of the work:

(a)   Status of work in progress and current projected completion dates.

(b)   Work completed since previous report including dates of completion.

(c)   Work rescheduled from contract program dates together with reasons therefore.

(d)   Work where current or anticipated delays will affect the component completion dates together with remedial actions taken by the Contractor.

15.3   DOCUMENT AMENDMENTS DURING BIDDING

15.3.1   Supplemental Notice

REDMAN CHEMICALS CORPORATION
MONTCLIFF AVENUE
REEDTOWN CA 00012

TELEPHONE: 035 478 925   FACSIMILE: 035 734 045

SUPPLEMENTAL NOTICE No. 1

Date of issue: March 20, 1992

To: ....
                    (*name of bidder*)

            ....
                    (*address of bidder*)

            ....

On the Work for:

Altona Plant Expansion Development Project No. 7780
Centrifugal Air Compressors—Contract No. B/417

the Bidding Documents are modified as follows:

1.A.   TECHNICAL REQUIREMENTS FORM—REVISION 1

1.B.   CENTRIFUGAL AIR COMPRESSOR SPECIFICATION SP-1503—REVISION 1

Approved By: K. T. Kenwood (Project Manager)

END OF SUPPLEMENTAL NOTICE No. 1

15.3.2 Document Transmittal Notice

REDMAN CHEMICALS CORPORATION
MONTCLIFF AVENUE
REEDTOWN CA 00012

TELEPHONE: 035 478 925   FACSIMILE: 035 734 045

DOCUMENT TRANSMITTAL NOTICE No 1

Date of issue: March 30, 1992

To: ....

Attention: ....

Transmitted herewith are documents for:

.( ) Approval   ( ) Per your request
( ) Preliminary   ( ) Onformation
( ) Reference only   ( ) Construction
( ) Revised as noted   (x) Quotation/bidding
( ) Approved as noted   ( ) Other ....

"x" where applicable

Document No.	Revision No.	P-print T-transp. O-other	Number of Copies	Document Title
	1	O	1	TECHNICAL REQUIREMENTS FORM
SP -1503	1	O	1	CENTRIFUGAL AIR COMPRESSOR SPECIFICATION

Please acknowledge receipt of this transmittal by returning one signed copy to the undersigned.

Issued by: K. T. Kenwood   *KTK.*

       Project Manager

Received by: .... Date: ....

15.3.3   Revised Technical Requirements Form

REDMAN CHEMICALS CORPORATION

TECHNICAL REQUIREMENTS FORM

TITLE Centrifugal Air Compressors	**CONTRACT No** B/417
	PROJECT No 7780
DELIVER TO Redman Chemicals Corporation	**REVISION No** *θ1*
Altona Plant	
Montcliff Avenue	**SHEET No** 1 **OF** 1
Reedtown CA 00012	

ATTACHMENTS

	THE FOLLOWING DOCUMENTS FORM PART OF THE TECHNICAL SPECIFICATION		REV
1	Drawing and Technical Data Submittal Requirements Form		0
2	Centrifugal Compressor Specification	SP − 1503	*θ1*
3	Standard General Data and Requirements Specification	SSP − 0201	1
4	* Standard Electric Motor Specification	SSP − 3302	3
5	* Standard Noise Specification	SSP − 5101	2
6	* Standard Painting Specification	SSP − 5603	5
7	Centrifugal Air Compressor Data Sheet	DS − 1503 − 1	0
8	Induction Motor Data Sheet	DS − 3302 − 1	0
9			
10			

REV	ITEM NO	QTY & UNIT	DESCRIPTION	ACCOUNT NO
1	1	*2 3*	Centrifugal air compressors with electric	
			motor drivers complete with all accessories	
			as specified	
1			Tag Equipment: 1501 and 1502 and 1503	7780/4
			* Assumed to be in Bidder's possession	

REV	DATE	ISSUE DESCRIPTION	BY	APPROVALS
1	Mar 18,92	Revised as noted and reissued for bids	TWW	*KTK*
0	Mar 7,92	Issued for bids	T. White	*KTK*

This section presents examples for the completion of in-house commercial and technical bid analysis sheets for the selection of a recommended contractor to undertake the supply and delivery of major plant equipment such as the centrifugal air compressors for this sample procurement.

15.4   BID EVALUATION AND NEGOTIATION

15.4.1   In-House Bid Analysis Sheets

IN-HOUSE BID ANALYSIS SHEETS
Centrifugal Air Compressors—Contract No. B/417

PREPARATION OF BID ANALYSIS

1. *Responsibility*
   The engineer who wrote the equipment specification and prepared the equipment data sheets (see Section 15.2) should be responsible for preparing the Bid Analysis including the technical analysis and analysis of pricing.

2. *Formats*
   The following set of forms should be referenced as practical guides to the preparation of the Bid Analysis for the sample procurement covering the supply and delivery of the centrifugal air compressors.

   - Bid Analysis Summary And Recommendation Sheet (1 sheet)
   - Technical Bid Analysis Sheets (4 sheets)

3. *General Requirements for Completing the Analysis Sheets*

   Possible errors in the bidder's offer should be sought. Quoted performance should be cross-checked with performance curves or other data that will aid in the understanding of the bidder's offer. As much information as necessary should be tabulated to make a real comparitive analysis, that is, enough to establish conclusively which equipment is best for the application and why it is so. It is also necessary to establish why other bids, especially less expensive ones, are not desirable or not acceptable.

   Notes:
   - (a) Where no bidder's details are available, show item as "included" (in Bid) or "not included"
   - (b) No prices or deliveries are to be shown in the technical analysis sheets
   - (c) The name of the recommended bidder is to be noted in the RECOMMENDED BID-DER fill-in space in the Bid Analysis Summary and Recommendation Sheet. The basis of the recommendation, that is, Lowest *evaluated* price that meets specifications, is to be entered under the Remarks space in the respective bidder's column.

REDMAN CHEMICALS CORPORATION

BID ANALYSIS SUMMARY AND RECOMMENDATION SHEET

EQUIPMENT   Centrifugal Air Compressors
EQUIPMENT TAG No.   1501, 1502, 1503
CONTRACT No.   B/417

	SPECIFIED	BIDDER A (Bid Ref: / Date:)	BIDDER B (Bid Ref: / Date:)	BIDDER C (Bid Ref: / Date:)	BIDDER D (Bid Ref: / Date:)	BIDDER E (Bid Ref: / Date:)
COSTS AND OTHER INFORMATION						
1 PRICE COMPONENTS – EX WORKS	Local Currency					
2 PRICE COMPONENTS – FAC, FOB, FAS	Local Currency					
3 EXCHANGE RATE (DATE)	To $ US					
4 EXCHANGE RATE (DATE)	To $ US					
5 TOTAL COST OF OVERSEAS COMPONENTS CONVERTED TO $ US						
6 TRANSPORTATION – OCEAN	$ US					
7 TRANSPORTATION – LAND	$ US					
8 INSURANCE	$ US					
9 CUSTOMS DUTY	$ US					
10 ON – SITE ADVISORS	$ US	1 for 15 days				
11 ESSENTIAL SPARE PARTS	$ US					
12 ADJUST FOR:	$ US					
13 ADJUST FOR:	$ US					
14 ADJUST FOR:	$ US					
15 TOTAL CONTRACT PRICE $ US						
16 FIXED OR SUBJECT TO ESCALATION	Fixed					
17 TERMS OF PAYMENT	FOB 90/5/5 %					
18 Cost comparison for power demand / energy consumption						
19						
20 COUNTRY OF ORIGIN						
21 DELIVERY TIME EX WORKS (WEEKS)						
22 DELIVERY TIME SITE (WEEKS)						
23 SHIPPING POINT						
RECOMMENDED BIDDER		REMARKS	REMARKS	REMARKS	REMARKS	REMARKS
RELEASED FOR PURCHASE	BUDGET ESTIMATE					
	$1,300,000					
	incl. spares					

BY	CHECKED	APP'D	DATE

REDMAN CHEMICALS CORPORATION
TECHNICAL BID ANALYSIS SHEET

EQUIPMENT Centrifugal Air Compressors
EQIPMENT TAG No. 1501, 1502, 1503
CONTRACT No. B/417

	DESCRIPTION	SPECIFIED	BIDDER A Bid Ref: Date: M	S	BIDDER B Bid Ref: Date: M	S	BIDDER C Bid Ref: Date: M	S	BIDDER D Bid Ref: Date: M	S	BIDDER E Bid Ref: Date: M	S
1	COMPRESSOR DETAILS											
2	General:											
3	Manufacturer											
4	Type	Integral gear										
5	Make/Model											
6	Performance:											
7	Inlet flow	250 m ^{3}/min										
8	Inlet/discharge press	kPa 99.2 A/860 g										
9	Inlet/discharge temp	30°C /										
10	kW/speed											
11	Number of stages											
12	Impeller tip speed:											
13	Stages 1 and 2											
14	Stages 3 and 4											
15	Estimated surge flow											
16	Power at coupling											
17	CONSTRUCTION/MATERIALS											
18	Casing:											
19	Type/split											
20	Material											
21	Design press/temp											
22	Impellers:											
23	Type/material											
24	Diameter each stage											
			REMARKS		REMARKS		REMARKS		REMARKS		REMARKS	

M S – Meets Specification Yes / No

DATE	BY	APPROVED

REDMAN CHEMICALS CORPORATION
TECHNICAL BID ANALYSIS SHEET

EQUIPMENT Centrifugal Air Compressors

EQIPMENT TAG No. 1501, 1502, 1503

CONTRACT No. B/417

	DESCRIPTION	SPECIFIED	BIDDER A Bid Ref: Date:	M	S	BIDDER B Bid Ref: Date:	M	S	BIDDER C Bid Ref: Date:	M	S	BIDDER D Bid Ref: Date:	M	S	BIDDER E Bid Ref: Date:	M	S
1	Shaft material																
2	Labyrinth type/material																
3	Shaft Seal type/medium																
4	Intercoolers:																
5	Tube material																
6	Number of tubes/stage																
7	Tube ID/OD/length																
8	C.W. flow per cooler																
9	Aftercooler:																
10	Tube material																
11	Number of tubes																
12	Tube ID/OD/length																
13	Cooling water flow																
14																	
15	Coupling make/type																
16	Radial Bearing type																
17	Thrust Bearing type																
18																	
19	LUBRICATION and SEAL OIL																
20	Reservoir:																
21	Retention																
22	Heater kW																
23	Main Lube Oil Pump:																
24	Make/type																

M S – Meets Specification Yes / No

		REMARKS	REMARKS	REMARKS	REMARKS	REMARKS

DATE	BY	APPROVED

REDMAN CHEMICALS CORPORATION

TECHNICAL BID ANALYSIS SHEET

EQUIPMENT Centrifugal Air Compressors

EQIPMENT TAG NO.   1501, 1502, 1503

CONTRACT No. B/417

DESCRIPTION	SPECIFIED	BIDDER A Bid Ref: Date:		BIDDER B Bid Ref: Date:		BIDDER C Bid Ref: Date:		BIDDER D Bid Ref: Date:		BIDDER E Bid Ref: Date:	
		M	S	M	S	M	S	M	S	M	S
1 Auxiliary Lube Oil Pump:											
2 Make/type/kW											
3 Lube Oil Filter:											
4 Make/type											
5 Lube Oil Cooler:											
6 Number/type											
7 Cooling water flow											
8 Lube & Seal Oil Piping:											
9 Material											
10											
11 ELECTRIC MOTOR											
12 Manufacturer/type											
13 Voltage/Phase/Hz											
14 Frame/power (kW)											
15 Speed (rpm)											
16 Service factor											
17 Type of enclosure											
18 Type of insulation											
19 Type of bearings											
20 Bearing lubrication											
21											
22 INSTRUMENTATION/CONTROLS											
23 Scope as per Compr. Spec.?											
24 Added equipment (if any)											
M S – Meets Specification Yes/No		REMARKS		REMARKS		REMARKS		REMARKS		REMARKS	

DATE	BY	APPROVED

REDMAN CHEMICALS CORPORATION
TECHNICAL BID ANALYSIS SHEET

EQUIPMENT Centrifugal Air Compressors
EQIPMENT TAG No. 1501, 1502, 1503
CONTRACT No. B/417

	DESCRIPTION	SPECIFIED	BIDDER A		BIDDER B		BIDDER C		BIDDER D		BIDDER E	
			Bid Ref:		Bid Ref:		Bid Ref:		Bid Ref:		Bid Ref:	
			Date:		Date:		Date:		Date:		Date:	
			M	S	M	S	M	S	M	S	M	S
1	ACCESSORIES											
2	Axial Probes:											
3	Type/make/model											
4	Vibration read-out:											
5	Type/make/model											
6	Vibration Monitor:											
7	Type/make/model											
8	Air Filter/Silencer:											
9	Type/make/model											
10	Blow-off silencer:											
11	Type/make											
12												
13	NOISE LIMITATION [dB(A)]											
14	Compressor at full load											
15	Blow-off silencer	max. 85 dB(A)										
16	Electric motor											
17												
18	TOTAL C.W. FLOW L/min											
19												
20	MASS (kg) & DIMENSIONS (mm)											
21	Compressor mass											
22	Compressor dimensions											
23	Compressor and driver mass											
24	Overall dimensions											
	M S − Meets Specification Yes / No		REMARKS		REMARKS		REMARKS		REMARKS		REMARKS	

DATE	BY	APPROVED

15.4.2   In-House Bid Evaluation Checklists

This section presents in-house bid evaluation checklists specifically designed for major plant equipment such as the sample procurement for the centrifugal air compressors. In practice, these checklists should be tailored to those factors that are relevant to the particular type of equipment and conditions under which it is procured.

IN-HOUSE BID EVALUATION CHECKLISTS
Centrifugal Air Compressors—Contract No. B/417

TECHNICAL

	Yes	No
TA. MAJOR TECHNICAL FEATURES		
1. Does the recommended bid generally conform with the specification(s)?	☐	☐
2. Are these primary requirements met, or information supplied?		
• power consumption	☐	☐
• cooling water requiremen	☐	☐
• fail-safe provisions	☐	☐
• noise, sound pressure level	☐	☐
• can be installed in the allotted area	☐	☐
• ease of operation	☐	☐
• ease of maintenance	☐	☐
• compatibility with existing equipment	☐	☐
3. Are the key compressor components interchangeable?	☐	☐
4. Is the equipment offered a proven product for reliability for the intended service?	☐	☐
5. Are alternate proposals (if submitted) realistic and acceptable?	☐	☐
6. Have estimated manhours for equipment installation, commissioning and personnel training by contractor been included in bid?	☐	☐
7. Have shipping mass (kg) and crate dimension details been listed?	☐	☐
TB. GENERAL TECHNICAL MATTERS		
1. Does the recommended bidder agree to adhere to the technical drawing and data submittal requirements?	☐	☐
2. Has the recommended bidder a complete understanding of the scope of the operating and maintenance manual requirements to be submitted?	☐	☐
3. Has a preliminary work program been submitted?	☐	☐
4. Has a quality assurance write-up been submitted?	☐	☐

IN-HOUSE BID EVALUATION CHECKLISTS (cont.)

	Yes	No
5. Are contractor and subcontractor manufacturing facilities ISO 9000 registered?	☐	☐
6. Are the nominated subcontractors to perform listed items of the work acceptable?	☐	☐
7. Are maintenance back-up facilities and availability of spare parts acceptable?	☐	☐
8. Has a list been included for items classified for normal operational wear and tear and are not subject to the equipment warranty (defects liability)?	☐	☐
9. Has a list of special tools been given?	☐	☐
10. Has a list of itemized priced spare parts been submitted to cover the specified period of time?	☐	☐

COMMERCIAL

	Yes	No
1. Has the recommended bid been signed by an authorized person?	☐	☐
2. Have all the supplemental notices been acknowledged?	☐	☐
3. Is the recommended bid conforming to commercial requirements?	☐	☐
4. Have all schedules been submitted?	☐	☐
5. Has the length of bid validity period been confirmed?	☐	☐
6. Are supplementary price schedules complete?	☐	☐
7. Have foreign currency details been supplied?	☐	☐
8. Have customs tariff items been listed?	☐	☐
9. Is the contract price fixed?	☐	☐
10. Is the contract price subject to escalation?	☐	☐
11. Is delivery program with deadlines satisfactory?	☐	☐
12. Is the contractor responsible for freight cost?	☐	☐
13. Does the recommended bidder agree with the terms of payment?		
(a) Down payment	☐	☐
(b) Interim payments	☐	☐
(c) Final payment	☐	☐
14. Does the recommended bidder agree with the specified liquidated damages?	☐	☐
15. Have taxes been included in the bid price?	☐	☐
16. Is insurance and risk of loss covered?	☐	☐
17. Are the alternative bids priced?	☐	☐

IN-HOUSE BID EVALUATION CHECKLISTS (cont.)

GENERAL

	Yes	No

1. Is completeness and thoroughness of the recommended
   bid acceptable? ☐ ☐
2. Would there be language or other communication problems? ☐ ☐
3. Has an assessment been made of other production commitments by the
   contractor which may prevent him meeting the contract program? ☐ ☐
4.[1] Have management skills and management involvement in the contract
   been identified? ☐ ☐
5.[1] Have a company organization chart and résumés of key personnel assigned to the contract and their responsibilities been submitted? ☐ ☐
6.[1] Has the following related experience been included?
   (a) "Boilerplate" company profile ☐ ☐
   (b) Financial statement ☐ ☐
   (c) List of similar installations—home country as well as overseas contracts ☐ ☐

---

[1] This information would normally not be required when soliciting bids from well-known specialist equipment suppliers. It is needed, however, for very large procurements such as a
power plant, coal handling system, and where equipment is solicited on a worldwide basis.

15.4.3 Minutes of Meeting Form—A Record of Negotiations

REDMAN CHEMICALS CORPORATION

WRITTEN CONFIRMATION

CONFIRMATION No 3   **CONTRACT No** B/417   **SHEET** 1 **of** 1

CONFERENCE / ~~TELEPHONE~~   **DATE** May 23, 1992

MEETING HELD AT : Redman Chemicals Head Office, Reedtown

SUBJECT : Negotiating outstanding contractual items

RECORDED BY :   Terry White

PRESENT :

Redman Chemicals	Beevor Compressor, Inc.
K.T. Kenwood	V. Owell
B.F. Thomas	J.P. Cassidy
T.W. White	

ACTION BY	
	The following were the salient items negotiated and agreed upon:
Beevor	**Item 1 Revised delivery** All items of equipment to be received at the Altona site no later than January 5, 1993.
	Item 2 Location of compressors and termination points Redman tabled Dwg 7780-001-2, Rev A (issued for comment) showing the proposed location of the equipment, and the Purchaser/Contractor termination points of supply. Beevor confirmed that there is adequate space available between the units for both operation and maintenance. Beevor also confirmed acceptance of the location of the termination points.
Beevor	**Item 3 Installation, commission and personnel training** Beevor will make available Mr. Bob Brown, Chief Service Engineer, for checking the installation and assisting with equipment start-up operations. Two weeks prior to delivery, Mr. Brown will conduct a series of training sessions for plant operators and maintenance personnel.

DISTRIBUTION :
K.T. Kenwood
L.P. Lechfield (Engineering)   Beevor   3 copies
H.I. Weeks (Construction)
P.T. Ouchmann (Production)
A.M. Jenkins (Purchasing)
File

15.5 CONTRACT NOTIFICATIONS

15.5.1 Notification of Award of Contract

Note: This document has important legal consequences. Consultation with an attorney is encouraged with respect to its wording and completion of content.

REDMAN CHEMICALS CORPORATION
MONTCLIFF AVENUE
REEDTOWN CA 00012

TELEPHONE: 035 478 925   FACSIMILE: 035 734 045

KTK/rg
Ref. B/7780/3/xh

June 15, 1992

Mr. Vincent Owell
Sales Executive
Beevor Compressor, Inc.
2418 Balany Drive
Tingory GJ 710000

Dear Mr. Owell:

Altona Plant Expansion Development Project No. 7780
Centrifugal Air Compressors—Contract No. B/417
NOTICE OF AWARD

You are advised that your revised offer, reference 721 - 4914 for execution of the above named contract for the supply and delivery of the three compressors for the fixed price of $................, and your listed schedule of rates for any additional work, has been accepted.

The date of contract is the date of this Notice of Award.

A certified copy of the proposed contract is attached hereto for your review and any items with which you are not in agreement should be evidenced in writing within five (5) days of this Notice of Award. If you are in agreement with the contract document please sign the duplicate copy of this notification and return by June 20, 1992. If this letter is not received by that date it will be deemed that you are in complete agreement.

A formal Contract Agreement evidencing the terms and conditions of the Contract will be prepared and submitted for your execution within the time stated in the Contract Conditions.

No work should commence until insurance policies are in place and a copy of the policy or evidence of payment of premium forwarded within ten (10) days of this Notice of Award.

Would you kindly acknowledge by signing the enclosed copy at the appropriate places and return it to us to be received no later than 4:00 pm, June 20, 1992.

Sincerely,
Redman Chemicals Corporation

....
K. T. Kenwood - Project Manager

We hereby acknowledge receipt of the Notice of Award dated June 15, 1992.

....
Duly authorized for and behalf of

....
Name of Contractor

Dated: ....

15.5.2   Notification of Bid Rejection

REDMAN CHEMICALS CORPORATION
MONTCLIFF AVENUE
REEDTOWN CA 00012

TELEPHONE: 035 478 925   FACSIMILE: 035 734 045

KTK/rg
Ref. B/7780/12/jt

June 25, 1992

....

   (*Bidder's company name*)

....

   (*Address*)

....

Attention: ....
                 Sales Manager

Altona Plant Expansion Development Project No. 7780
Centrifugal Air Compressors—Contract No. B/417

Gentlemen:

With reference to your Bid (*reference number*) dated .... for the above contract, we regret to advise you that your offer has been unsuccessful.

We thank you for the interest you have taken with the preparation and presentation of your submission and assure you that we will seek bid enquiries from your Company for similar work in future.

Sincerely,
Redman Chemicals Corporation

K. T. Kenwood - Project Manager

LOCATOR GUIDE FOR KEY SUBJECT MATTER IN BIDDING AND CONTRACT DOCUMENTS

This Appendix presents a locator guide for the uniform location of subject matter in bidding and contract documents for major plant equipment procurements. The purpose of the guide is to assist anyone confronted with the preparation of these documents, and who is faced with numerous decisions as to which of the various documents involved should deal with a particular topic, and where a particular paragraph is to be located. The uniform locator guide will also benefit those who work in different company departments when preparing technical and nontechnical documents for a particular procurement. Uncoordinated efforts in these circumstances can result in conflicting statements and requirements. Furthermore, uncertainties as to where to locate a topic has frequently led to repetition of the topic in various documents (sometimes inevitable), and produced contradictions and confusion among the users of the documents, often with unanticipated legal consequences. This locator guide will remedy these situations.

The locator guide contains a tabulated listing of most key topics that are commonly included in any of the following bidding (instructions for bidding), commercial or technical documents that have been discussed in this book:

1. Instructions to Bidders
2. General (Standard) Contract Conditions
3. Supplementary Conditions
4. Design and Performance Specification
5. General Data and Requirements Specification

As discussed in Chapter 3, certain bidding documents do not become contract documents when formulating the contract. In Chapter 5, the contract conditions were separated to cover those conditions which are typical to most equipment procurements, General (Standard) Conditions, and those that apply to a particular procurement, Supplementary Conditions.

Chapter 7 made reference to the technical design and performance requirement for a particular equipment procurement to be specified in the Design and Performance

Specification, and general technical data and requirement provisions that would apply to most of the procurements undertaken by a company, to be located in an in-house standard specification titled General Data and Requirements Specification.

The locator guide is also intended to discourage the frequent practice of covering the same topic in two or more documents. To cover a certain technical or commercial topic once and clearly should be sufficient. Some subject matter such as contract administration procedures, and work-related issues (other than technical scope of work) applying to a particular contract, may present difficulties in deciding whether the topic is to be located in a technical or commercial document. There are exceptions where repetition has to be utilized for emphasis and clarity. Much will depend on who will be the users of the document. Wherever possible, "say it once," but "say it correctly."

SUBJECT MATTER	DOCUMENT
Acceptance of the Work - definition	General Contract Conditions
Acceptance of the Work - procedure	Supplementary Conditions
Administrative Procedures:	
Application for Payment	Supplementary Conditions
Application for Final Payment	Supplementary Conditions
Certificates - application for	Supplementary Conditions
Change Order	Supplementary Conditions
Payment Claims - approval	Supplementary Conditions
Progress Reporting	Supplementary Conditions, *or* General Data and Requirements Specification
Taking Over the Work	Supplementary Conditions
Work Program	Supplementary Conditions, *or* General Data and Requirements Specification
Alternative Bids	Instructions to Bidders
Arbitration - general	General Contract Conditions
Arbitration - applicable to the contract	Supplementary Conditions
Award - basis for bidding	Instructions to Bidders
Award - notice of	Instructions to Bidders
Bid - number of copies required	Instructions to Bidders
Bid Evaluation - basis for bidding	Instructions to Bidders
Bid Form	Instructions to Bidders
Bid Opening - public or private	Instructions to Bidders
Bid Security Requirement	Instructions to Bidders
Bidding Documents - listing	Instructions to Bidders
Bidding Documents - clarification	Instructions to Bidders
Bidding Period	Instructions to Bidders
Bonds - contract security requirements	Supplementary Conditions
Certificates - general	General Contract Conditions

SUBJECT MATTER	DOCUMENT
Certificates - specific requirements	Supplementary Conditions
Codes, Standards, and Regulations	Design and Performance Specification
Communications prior to Bidding Closing	Instructions to Bidders
Copies of Bid - number required	Instructions to Bidders
Contract Agreement - execution of	Instructions to Bidders / General Conditions
Contract Documents - general	General Contract Conditions
Contract Documents - clarification	Instructions to Bidders
Contract Price - for bidding	Instructions to Bidders
Contract Price - specific	Supplementary Conditions
Contract Price Adjustment - escalation	Supplementary Conditions
Contract Program	Supplementary Conditions
Correction of Defective Work	Design and Performance Specification
Defects Liability - general	General Contract Conditions
Defects Liability - specific technical	Design and Performance Specification
Delivery - time for	Supplementary Conditions
Delivery - location	Supplementary Conditions
Drawings & Data - use by others	General Contract Conditions
Drawings & Data - issued by purchaser	Design and Performance Specification
Drawings & Data - submittal requirements	General Data and Requirements Specification
	Drawing & Technical Data Submittal Form
Equipment Warranty - definition (see also Defects Liability)	General Contract Conditions
Equipment Warranty - technical defects	Design and Performance Specification
Equipment Warranty - performance	Design and Performance Specification
Extension of Time	General Contract Conditions
Inspection - general	General Contract Conditions
Inspection - specific requirements	General Data and Requirements Specification Equipment Data Sheet
Insurance - general	General Contract Conditions
Insurance - specific	Supplementary Conditions
Late Bids	Instructions to Bidders
Liquidated Damages - general	General Contract Conditions
Liquidated Damages-specific requirements	Supplementary Conditions
Modifications, Amendments to Bidding Documents	Instructions to Bidders
Modifications, Withdrawals of Bid	Instructions to Bidders
Nonconforming Bid	Instructions to Bidders
Omissions, Errors, or Discrepancies	General Contract Conditions
Operating and Maintenance Manuals	General Data and Requirements Specification
	Drawing & Technical Data Submittal Form
Performance Test	Design and Performance Specification

SUBJECT MATTER	DOCUMENT
Permits	General Contract Conditions
Price Escalation	Supplementary Conditions
Rejecting Bids	Instructions to Bidders
Rejecting Defective (Faulty) Work	General Contract Conditions
Retention Money	Supplementary Conditions
Shipping - general	Supplementary Conditions
Shipping - specific assembly and crating	General Data and Requirements Specification
Scope of Work - technical	Design and Performance Specification
Spare Parts - specific requirement	Design and Performance Specification
Spare Parts - general requirements	General Data and Requirements Specification
Taking Over the Work - definition	General Contract Conditions
Taxes - sales and VAT	Supplementary Conditions
Tests - shop	General Data and Requirements Specification
	Equipment Data Sheet
Type of Contract	Instructions to Bidders
Validity Period	Instructions to Bidders
Variations - general	General Contract Conditions
Warranty - general	General Contract Conditions
Work by the Contractor	Design and Performance Specification
Work by the Purchaser (or others)	Design and Performance Specification

SI UNITS AND CONVERSIONS

SI is an abbreviation for "Système International d'Unités" (International System of Units) adopted by the Eleventh General Conference on Weights and Measurements in 1960. It is a system of internationally recognized metric units of measurement employing:

- Base units of SI—Table 1
- Supplementary units of SI—Table 2
- Derived SI units having special names—Table 3
- Decimal multiples and submultiples of SI units formed by use of prefixes—Table 4

This Appendix also contains a list of common SI units used in engineering—Table 5, and conversions to SI units—Tables 6 and 7.

Base Units. The base units together with their symbols are shown in the table hereunder.

TABLE 1   Base Units of SI

Quantity	Name of Unit	Symbol	
length	meter	m	
mass	kilogram	kg	(see Note 1)
time	second	s	
electric current	ampere	A	
thermodynamic temperature	kelvin	K	(see Note 2)
luminous intensity	candela	cd	
amount of substance	mole	mol	(see Note 3)

Notes:

1. It should be noted that the unit of mass is the kilogram (1000 gram) and not the gram.
2. It should also be noted that the unit of kelvin is the unit of *Absolute* temperature. For most practical purposes, the degree Celsius is used. Note that centigrade is 1/100 of a right angle and should not be used for temperature. Note that the Kelvin and the degree Celsius are the same size but the kelvin scale has its zero at minus 273.15° C. The derived SI unit for temperature is shown in Table 3.

3. Where the mole is used, the elementary entities must be specified and may be atoms, molecules, ions, electrons, other particles or specified groups of particles.

Supplementary Units. In the SI system, the quantities plane angle and solid angle are treated as independent quantities with the units of radian and steradian respectively. They are described as supplementary units as shown in the table hereunder.

TABLE 2   Supplementary Units of SI

Quantity	Name of Unit	Symbol
plane angle	radian	rad
solid angle	steradian	sr

Derived Units. Expressions for the coherent derived units of the SI are stated in terms of the base units, e.g., the SI unit for density is kilogram per cubic meter (kg/m^3).

Some derived SI units have been assigned special names as shown in the table hereunder.

TABLE 3   Derived SI Units Having Special Names

Physical Quantity	Name of Unit	Symbol	Derivation
frequency	hertz	Hz	$1/s$
force	newton	N	$kg{\cdot}m/s^2$
pressure and stress	pascal	Pa	N/m^2
work, energy, quantity of heat, and impact strength	joule	J	$N{\cdot}m$ (see Note)
power	watt	W	J/s
electrical charge	coulomb	C	$A{\cdot}s$
electric potential	volt	V	W/A
electric capacitance	farad	F	C/V
electric resistance	ohm	Ω	V/A
electrical conductance	siemens	S	A/V
magnetic flux	weber	Wb	$V{\cdot}s$
magnetic flux density	tesla	T	Wb/m^2
inductance	henry	H	$V{\cdot}s/A$

TABLE 3   Derived SI Units Having Special Names (cont)

Physical Quantity	Name of Unit	Symbol	Derivation
temperature	degree Celsius	°C	For temperature interval $1\,°C = 1\,K$ For temperature $°C = K -273.15$
luminous flux	lumen	lm	cd•sr
illuminance	lux	lx	lm/m^2

Note: The energy required (or work expended) to break a specimen is measured in joules; to give a measure of impact resistance or strength.

Prefixes—Multiples and Submultiples of Units. Multiples and submultiples of units used in the metric system are named by attaching one of the following prefixes to the name of the unit e.g. kilometer (km), and megawatt (MW).

SI units should be used to indicate orders of magnitude, thus eliminating insignificant digits and decimals as well as providing a convenient substitute for writing powers of 10 (standard notation) which are generally preferred in computation.

The prefixes are shown in the table hereunder.

TABLE 4   Prefixes—Decimal Multiples and Submultiples

Prefix	Symbol	Multiples and Submultiples	Amount
mea	M	10^6	1 000 000
kilo	k	10^3	1 000
hecto	h	10^2	100
deca	da	10^1	10
deci	d	10^{-1}	0.1
centi	c	10^{-2}	0.01
milli	m	10^{-3}	0.001
microg	μ	10^{-6}	0.000 001
nano	n	10^{-9}	0.000 000 001

TABLE 5   Common SI Units Used in Engineering

Quantity	Name of Unit	Symbol	
Space and Time			
plane angle	radian	rad	
solid angle	steradian	sr	
length	meter	m	
area	square meter	m^2	
volume	cubic meter	m^3	
time	second	s	
angular velocity	radian per second	rad/s	
angular acceleration	radian per second squared	rad/s^2	
velocity	meter per second	m/s	
acceleration	meter per second squared	m/s^2	
flow rate (volume basis)	cubic meter per second	m^3/s	
Periodic and Related Phenomena			
frequency	hertz	Hz	
rotational frequency	revolution per second	1/s	(see Note 1)
	radian per second	rad/s	
Mechanics			
mass	kilogram	kg	
density	kilogram per cubic meter	kg/m^3	
momentum	kilogram-meter per second	kg•m/s	
moment of momentum, and angular momentum	kilogram-square meter per second	$kg•m^2/s$	
moment of inertia	kilogram-square meter	$kg•m^2$	
force	newton	N	
moment of force (torque)	newton-meter	N•m	
pressure and stress	pascal	Pa	
viscosity (dynamic)	pascal-second	Pa•s	(see Note 2)
viscosity (kinematic)	square meter per second	m^2/s	
surface tension	newton per meter	N/m	
energy, work, impact strength	joule	J	
power	watt	W	
Heat			
thermodynamic temperature	kelvin	K	
temperature interval	degree Celsius	°C	

TABLE 5   Common SI Units Used in Engineering (cont)

Quantity	Name of Unit	Symbol
Heat (cont)		
linear expansion coefficient	meter per meter per kelvin	K^{-1}
quantity of heat	joule	J
heat flow rate	watt	W
intensity of heat flow rate	watt per square meter	W/m^2
thermal conductivity	watt per meter-kelvin	$W/(m{\cdot}K)$
heat transfer coefficient	watt per square meter-kelvin	$W/(m^2{\cdot}K)$
heat capacity	joule per kelvin	J/K
specific heat capacity	joule per kilogram-kelvin	$J/(kg{\cdot}K)$
specific energy	joule per kilogram	J/kg
specific enthalpy	joule per kilogram	J/kg
entropy	joule per kelvin	J/K
specific entropy	joule per kelvin-kilogram	$J/(K{\cdot}kg)$
Electricity and Magnetism		
electric current	ampere	A
electric charge	coulomb	C
volume density of charge	coulomb per cubic meter	C/m^3
surface density of charge	coulomb per square meter	C/m^2
electric field strength	volt per meter	V/m
electric potential	volt	V
capacitance	farad	F
current density	ampere per square meter	A/m^2
magnetic field strength	ampere per meter	A/m
magnetic flux density	tesla	T
magnetic flux	weber	Wb
self-inductance	henry	H
permeability	henry per meter	H/m
magnetization	ampere per meter	A/m
resistance, impedance	ohm	Ω
conductance	siemens	S
resistivity	ohm-meter	Ω·m
conductivity	siemens per meter	S/m
reluctance	1 per henry	1/H

TABLE 5   Common SI Units Used in Engineering (cont)

Quantity	Name of Unit	Symbol
Light		
luminous intensity	candela	cd
luminous flux	lumen	lm
illumination	lux	lx
luminance	candela per square meter	cd/m^2
Acoustics		
period	second	s
frequency	hertz	Hz
wavelength	meter	m
density	kilogram per cubic meter	kg/m^3
sound pressure	pascal	Pa
sound particle velocity	meter per second	m/s
volume velocity	cubic meter per second	m^3/s
velocity of sound	meter per second	m/s
sound power	watt	W
sound intensity	watt per square meter	W/m^2
specific acoustic impedence	pascal-second per meter	Pa•s/m
mechanical impedence	newton-second per meter	N•s/m

Notes:

1. Rotational frequency may also be expressed r/s (revolutions per second)
2. While the SI unit for dynamic viscosity is Pa•s, it is not yet in widespread use. In the engineering of chemical plant and the specification of major equipment, the CGS unit the Poise (P), or more frequently its submultiple the centpoise (cP), remains the most common measure of dynamic viscosity, but it should not be used together with SI units. Note that $1\ cP = 10^{-2}\ P$

TABLE 6   Approximate Unit Conversions—Imperial to SI Units

Quantity	Imperial	SI
length	1 in	25.4 mm exact
	1 ft	0.3048 m
	39.37 in	1 m
	3.28 ft	1 m
area	1000 sq in	0.645 m^2
	10 sq ft	0.929 m^2
	1550 sq in	1 m^2
	10.76 sq ft	1 m^2
mass	1 lb	0.454 kg
	2.205 lb	1 kg = 1000 gram
	2205 lb	1 tonne = 1000 kg
	1 U.S. short ton	0.9072 tonne
	1 U.S. long ton	1.0161 tonne
	1 Imperial ton	1.0161 tonne
volume (see Note)	1 cu in	1.639×10^{-5} m^3
	1 cu ft	0.0283 m^3
	35.315 cu ft	1 m^3
pressure	1 lb/sq in	6.895 kPa
	0.145 lb/sq in	1 kPa
heat energy	1 Btu	1.055 kJ
	0.9478 Btu	1 kJ
heat flow rate	1 Btu/hr	0.293 W
	1 W	3.412 Btu/h
heat per unit mass	1 Btu/lb	2.326 kJ/kg
	1 kJ/kg	0.4299 Btu/lb
specific heat capacity	1 Btu/lb °F	4.1878 kJ/kg•°C
	1 kJ/kg•°C	0.239 Btu/lb °F
density of heat flow rate	1 Btu/ft^2 hr	3.155 W/m^2
	1 W/m^2	0.317 Btu/ft^2 hr
heat transfer coefficent	1 Btu/ft^2 hr °F	5.678 W/m^2•°C
	1 W/m2•°C	0.176 Btu/ft2 hr °F

Note: The liter is also used as a unit of volume and is equal to 1000 cm^3 exactly. It is specifically used to express volumes of fluids. The cubic meter is used where solids and gases and actual physical spaces are being considered.

TABLE 7   Conversion Factors

	To convert	Multiply by	To obtain SI unit
acceleration	ft/sec^2	0.3048	m/s^2
area	in^2	645.16	mm^2
	in^2	6.452×10^{-4}	m^2
	ft^2	0.0929	m^2
density	$slug/ft^3$	515.4	kg/m^3
	lbm/ft^3	16.026	kg/m^3
energy (work	ft-lbf	1.356	J
or quantity	ft-lbf	3.77×10^{-7}	kWh
of heat)	Btu = 778.2 ft-lbf	1055.1	J
flow rate	lbm/sec	0.4536	kg/s
(mass)	lbm/min	7.599×10^{-3}	kg/s
	lbm/hr	1.260×10^{-4}	kg/s
flow rate	ft^3/sec	0.0283	m^3/s
(volumetric)	ft^3/min	0.0472×10^{-2}	m^3/s
	U.S. gpm	0.0631×10^{-3}	m^3/s
	Imp. gpm	0.0758×10^{-3}	m^3/s
force	lbf	4.448	N
heat transfer coefficient	$Btu/hr\text{-}ft^2\text{-}°R$	5.678	$W/(m^2 \bullet K)$
length	in	25.4	mm
	ft	0.3048	m
	mile	1.609	km
mass	slug	14.59	kg
	lbm	0.4536	kg
power	ft-lbf/sec	1.356	W
	hp = 550 ft-lbf/sec	745.7	W
pressure	$psi\ (lb/in^2)$	6,895	Pa
	lb/ft^2	47.88	N/m^2

TABLE 7   Conversion Factors (cont)

	To convert	Multiply by	To obtain SI unit
specific heat capacity	ft-lb/(slug-°R)	0.1672	J/(kg•K)
	Btu/lbm-°R	4,186	J/(kg•K)
specific volume	ft^3/lbm	0.06242	m^3/kg
specific weight	lbf/ft^3	157.09	N/m^3
torque	lbf-in	0.1130	N•m
	lbf-ft	1.3559	N•m
velocity	ft/sec	0.3048	m/s
	ft/min	0.0051	m/s
	mph	1.6093	km/h
viscosity (dynamics)	centipoise (cP)	10^{-3}	Pa•s
	lb-s/ft^2	47.88	Pa•s
viscosity (kinematic)	centistokes (cSt)	10^{-6}	m^2/s
	ft^2/sec	0.0929	m^2/s
volume	in^3	1.629×10^{-5}	m^3
	ft^3	0.0283	m^3
	U.S. gallon	0.00378	m^3
	Imp. gallon	0.00455	m^3

Note: °R is the absolute temperature in degrees F, i.e. degrees Rankine. For example, 0°F = 460°R.

TRADE TERMS

The trade terms offered by a bidder in a submitted bid for the supply of materials and/or equipment should always be defined by reference to a standard definition to qualify the condition of sale. Similar to the conditions of a procurement contract discussed in Chapters 5 and 6, there are conditions that set out the rights and obligations of each party when it comes to transporting the material and/or goods hereafter termed "goods." They define such situations as:

- The party responsible for arranging and paying for the transport (carriage) of the goods from one location to another.
- The party who will bear the costs and risks if these operations are not carried out successfully.
- The party who will bear the risk of loss of or damage of the goods in transit.

These conditions are called "trade terms." For the purpose of uniformity, trade terms used in this book for both overseas and domestic procurements, are Incoterms set out by the International Chamber of Commerce (ICC).[1] Versions of the Incoterms have been updated to take into account changes in the method of transportation (door-to-door inter-modal, containerization, roll-on/roll-off, etc) as well as EDP methods for expediting documents.

The thirteen Incoterms (1990 version) listed in increasing order of responsibility on the contractor (seller/supplier) are:

INCOTERMS—ABBREVIATION, GROUP AND TERM USED

Group E   Departure—Permits the contractor to make the goods available to the purchaser at the manufacturer's or supplier's premises stipulated in the contract.

EXW   Ex Works .... named place

---

[1]A copy of Incoterms 1990 is available from the ICC PUBLISHING, INC., 156 Fifth Avenue, Suite 820, New York, NY 10010, or from ICC National Committees or Councils.

Group F   Main Carriage Unpaid—Requires the contractor to carry the cost and risk, and deliver the goods to a carrier stipulated in the contract by the purchaser at the:

FCA   Free Carrier   ...... named place

FAS   Free Alongside Ship   ...... named port of shipment

FOB   Free on Board Ship   ...... named port of shipment

Group C   Main Carriage Paid—Stipulates that the contractor contracts for carriage, but does so without bearing the risk of loss or damage to the goods due to events arising after shipment or dispatch of the goods at the:

CFR   Cost and Freight   ..... named port of destination

CIF   Cost, Insurance, and Freight   ...... named port of destination

CPT   Carriage Paid To   ...... named place of destination

CIP   Carriage and Insurance Paid To   ...... named place of destination

Group D   Arrival—Requires the contractor to bear all the costs and risks needed to deliver the goods to the purchaser's stipulated location at the:

DAF   Delivered to Frontier   ...... named point and place

DES   Delivered Ex Ship   ...... named port of destination

DEQ   Delivered Ex Quay   ...... named port of destination

DDU   Delivered Duty Unpaid   ...... named place of destination

DDP   Delivered Duty Paid   ...... named place of destination

MODE OF TRANSPORT AND INCOTERMS

1. Any mode of transport including multimodal:
   EXW, FCA, CPT, CIP, DAF, DDU, and DDP

2. Air and rail transport:
   FCA

3. Sea and inland waterway transport:
   FAS, FOB, CFR, CIF, DES, and DEQ

INCOTERMS DEFINING DUTIES AND RESPONSIBILITIES

The following Incoterms define the duties and responsibilities of the contractor (seller/supplier) and the purchaser (buyer).

EXW—Ex Works ...... named place

Contractor has fulfilled his obligation to deliver when he has:
- Placed the goods at the disposal of the purchaser as provided in the contract for loading on the conveyance to be provided by the purchaser.

Purchaser:
- Bears all the costs and risks involved in loading and taking the goods from the contractor's or manufacturer's premises to the desired destination.

FCA—Free Carrier ...... named place

Contractor has fulfilled his obligation when he has:
- Delivered and handed over the goods into the charge of the carrier named by the purchaser in the contract.
- Cleared the goods for export.

The "Carrier" means any person or organization who, in a contract of carriage, undertakes to perform or procure carriage by rail, road, sea, air, inland waterway or any combination of such modes of transportation.

FAS—Free Alongside Ship ...... named port of shipment

Contractor has fulfilled his obligation when he has:
- Delivered the goods alongside the vessel pier or barge at the loading berth named by the purchaser at the port of shipment stipulated in the contract.
- Borne all costs and risks of the goods until such time as they have been effectively delivered alongside the vessel at the named port of shipment including any formalities required in order to deliver the goods alongside the vessel.
- Provided the purchaser with documents evidencing the delivery of the goods alongside the named vessel at the loading berth.

Purchaser:
- Bears all the charges and risks of loss or damage from the time when the goods have been effectively delivered alongside the vessel.
- Is responsible for clearing the goods for export.
- Bears any additional costs should the vessel named by purchaser fail to arrive on time or for any other cause after the contractor has placed the goods at the disposal of the purchaser at the date or within the period stipulated in the contract.

FOB—Free On Board ...... named port of shipment

Contractor has fulfilled his obligation when he has:
- Delivered the goods on board the vessel named by the purchaser at the named port of shipment and notified the purchaser accordingly.
- Borne all costs and risks of the goods until such time as they shall have effectively passed the ship's rail at the named port of shipment including any fees, taxes, and other costs in order to fulfill the loading of the goods on board the vessel.
- Cleared the goods for export.
- Provided the purchaser with documents evidencing the delivery of the goods on board the named vessel.

Purchaser:
- Undertakes all necessary arrangements for the booking of space on board a vessel, gives the contractor due notice of the name of, the vessel, loading berth of, and delivery dates to the vessel.
- Bears all costs and risks of loss or damage to the goods once they have passed the ship's rail.
- Bears any additional costs incurred because of the vessel having failed to arrive on the stipulated date or by the end of the period specified.

CFR—Cost and Freight ...... named port of destination

Contractor's obligations cover:
- Bearing all costs and freight necessary to bring the goods to the named port of destination stipulated in the contract.
- Responsibility for clearing the goods for export.
- Providing the purchaser with documents evidencing the delivery of the goods at the named port of destination.

Purchaser:
- Bears all costs and risks of loss or damage to the goods once they have been delivered to the named port of destination.

CIF—Cost, Insurance, Freight ...... named port of destination

Contractor's obligations cover:
- Paying for the carriage of goods to the agreed destination as stipulated in the contract by the usual route in a seagoing vessel.
- Paying for freight charges and unloading at the port of destination.
- Paying for the insurance premium for the risk of loss or damage to the goods during carriage.

Purchaser:
- Receives the goods at the agreed port of destination.
- Bears all risks of the goods from the time when they shall have effectively passed the ship's rail at the port of shipment. The insurance covers this to the port of destination.
- Pays all import duties as well as taxes payable at the time of or by reason of the importation.
- Procures and provides at his own risk and expense any import license and permit that may be required for the importation of the goods at the port of destination.

CPT—Carriage Paid To ...... named place of destination

Contract's obligations cover:
- Paying the freight for carriage of the goods to the named place of destination as stipulated in the contract.

Purchaser:
- Bears the risk of loss or damage to the goods and any additional costs after the goods have been delivered into the custody of the carrier at the named place of destination.

CIP—Carriage and Insurance Paid ...... named place of destination

Contract's obligations cover:
- Paying the freight and insurance for carriage of the goods to the named place of destination as stipulated in the contract.

Purchaser:
- Bears the risks of loss or damage to the goods and any additional costs after the goods have been delivered at the named place of destination.

DAF—Delivered at Frontier ...... named point and place

Contractor fulfilled his obligation when he has:
- Delivered the goods and they become available to the purchaser at the named point and place at the frontier, but before the customs border of the adjoining country, all as stipulated in the contract.
- Obtained all necessary documentation for exporting the goods.
- Borne all the costs and risks in delivering the goods to the named point and place at the frontier.

Purchaser:
- Obtains the necessary import documentation, pays any customs duty, taxes and other charges necessary for the importation of the goods.
- Takes possession of the goods as soon as they have been placed at his disposal at the named point and place at the frontier.
- Bears all risks of loss and damage to the goods from the time they have been placed at his disposal.

DES—Delivered Ex Ship ...... named port of destination

Contractor has fulfilled his obligation when he has:
- Delivered the goods and they become available to the purchaser on board the ship (uncleared) at the named port of destination as stipulated in the contract.
- Obtained all the necessary documentation for exporting the goods.
- Borne all the costs and risks in delivering the goods to the named port of destination.

Purchaser:
- Obtains the necessary import documentation, pays any customs duty, taxes and other charges necessary for the importation of the goods.
- Takes possession of the goods as soon as they have been placed at his disposal on board the vessel, and bears the costs of unloading them from the ship.
- Bears all risks of loss and damage to the goods from the time they have been placed at his disposal.

DEQ—Delivered Ex Quay (or Wharf), Duty Paid ...... named port of destination

Contractor has fulfilled his obligation when he has:
- Delivered the goods andy they become available to the purchaser on the quay or wharf at the named port of destination stipulated in the contract.
- Obtained all the necessary documentation for exporting and importing the goods.
- Borne all costs and risks, and paid import duties, taxes, and other charges of delivering the goods to the named port of destination.

Purchaser:
- Takes possession of the goods as soon as they have been placed at his disposal on the wharf at the named port of destination.
- Bears all risks of loss and damage to the goods from the time they have been placed at his disposal.

DDU—Delivered Duty Unpaid ...... named place of destination

Contractorr has fulfilled his obligation when he has:
- Delivered the goods and they become available to the purchaser at the named place of destination as stipulated in the contract.
- Obtained all the necessary documentation for exporting the goods.
- Borne all the costs and risks in delivering the goods to the named place of destination.

Purchaser:
- Bears costs for import duties, taxes and other charges payable upon importation.
- Obtains the necessary import documentation, pays any customs duty, taxes and other charges necessary for the importation of the goods.
- Bears all risks of loss and damage to the goods from the time they have been placed at his disposal.

DDP—Delivered, Duty Paid ...... named place of destination

Contractor has fulfilled his obligation when he has:
- Delivered the goods and they become available to the purchaser at the named place in the country of importation as stipulated in the contract.
- Obtained all necessary documentation for exporting and importing the goods.

- Borne all the costs and risks, and paid import duties, taxes, and other charges of delivering the goods to the named place of destination.

Purchaser:
- Takes possession of the goods as soon as they have been placed at his disposal.
- Bears all risks of loss and damage to the goods from the time they have been placed at his disposal.

Note that while the Incoterm EXW represents the minimum obligation for the contractor, DDP represents the maximum.

COMMON ABBREVIATIONS AND UNIT SYMBOLS USED IN TECHNICAL DOCUMENTS

This Appendix is concerned with the more common abbreviations and unit symbols used when writing engineering documents. Since there is no agreed-upon international standard for abbreviations, the following are used throughout this book. The systems are presented in order to achieve consistency between all bidding and contract documents as the same abbreviations often appear in more than one document. With the exception of engineering specifications, equipment data sheets and technical drawings, the use of abbreviations in written documents should be kept to a minimum.

This Appendix uses abbreviations for the International System of Units—SI (Le Système International d'Unités).

The following are some of the more commonly used abbreviations:

Time of day

These abbreviations are usually set in small caps

AM   ante meridiem (before noon)
PM   post meridiem (after noon)

10:50 AM   2:45 PM

Days of the week

Sun Mon Tues Wed Thur Fri Sat

Months of the year

Jan Feb Mar Apr May Jun Jul Aug Sep Oct Nov Dec

Date

The American style for dates differs from the British

American	month/day/year	December 10, 1995
British	day/month/year	10 December, 1995

It is no longer conventional practice to add a suffix to the number of the day, for example, 10th.

To avoid confusion in contract documents, the name of the month should be written and not referred to by a number. Avoid presenting a date as 12/10/1995. Internationally, this date can be interpreted to infer either December 10 or 12 October 1995, use:

12 October 1995, *or*
12 Oct '95

Points of the compass

Do not use periods when presenting points of the compass:

north N,   east E,   south S,   west W
north east NE,   north north east NNE

Units of measurement

Standard symbols and abbreviations for units of measurement should not be presented with periods:

kg   mm   kPa   MW   km/h

There should be one space between the number and the unit of measurement:

65 m   785 mm   94 kg   300 MW   842 kPa   26 °C

Abbreviations derived from foreign words

The following abbreviations are mainly of Latin origin:

e.g.	for example
i.e.	that is
etc.	and so on
vs.	versus

It should be noted that while the term *etc.* has been listed above, it should not be used in contract documents as it is too vague.

Abbreviations and unit symbols used mainly in engineering specifications, equipment data sheets, and technical drawings:

	Abbreviation	Unit Symbol
abbreviate	abbr	
absolute	abs	
above sea level	ASL	
alternating current	ac	
altitude	alt	
ambient	amb	
ampere		A
and	& (or spell out)	
appendix	appx	
approved	appd	
approximate	approx	
atmosphere	atm	
atmospheric (pressure)	atmos	
atomic weight	at wt	
auxiliary	aux	
average	avg	
barrel	bbl	
bearing	brg	
between centers	bc	
boiling point	bp	
Brinell hardness number	Bhn	
calorie		cal
calorific value	CV	
center line	cl	
center of gravity	cg	
center(s)	ctr(s)	
centimeter		cm
centimeter-gram-second (system)	cgs	
checked	ckd	
circular	circ	
coefficient	coef	
column	col	

(continued)

	Abbreviation	Unit Symbol
concentrated	conc	
conductivity	cndct	
constant	const	
construction	constr	
continuous - continued	cont	
cubic		cu
cubic centimeter		cc or cm^3
cubic meter		m^3
cubic millimeter		mm^3
cycles per second (Hertz)		cps or (Hz)
day	(spell out)	
decibel		dB
definition	def	
degree (angular)	deg	°
degree Celsius		°C
degree absolute (Kelvin)		°K
department	dept	
diagonal	diag	
diagram	diag	
diameter	dia	
direct current	dc	
document	doc	
drawing	dwg	
dyne	(spell out)	
each	ea	
efficiency	eff	
electric	elec	
electronic data processing	EDP	
elevation	el	
engineer	engr	
estimate	est	
equivalent	equiv	
exponential	expnt	
farad		F
figure	fig	

	Abbreviation	Unit Symbol
fluid	fl	
freezing point	fp	
frequency		Hz
full load	flld	
galvanize	galv	
gas	(spell out)	
gage (gauge)	ga	
government	govt	
grain	(spell out)	
gram	gr	g
gravitational acceleration		g
greater than	>	
hectare	ha	
height	hgt	
Hertz (cycles per second)		Hz
high pressure	hp	
high voltage	HV	
horizontal	horiz	
hour	(spell out)	h
hydrogen exponent	pH	
include	incl	
inside diameter	ID	
instrument	instr	
International System Units	SI	
joule		J
joule per kilogram		J/kg
joule per square meter		J/m^3
kilo (thousand)		k
kilocalorie		kcal
kilocycles per second		kHz/s
kilogram		kg
kilograms per cubic meter		kg/m^3
kilojoule		kJ

(continued)

	Abbreviation	Unit Symbol
kilojoule per kilogram		kJ/kg
kilometer		km
kilometers per second		km/s
kilometers per hour		km/h
kilonewton		kN
kilopascal		kPa
kilopascal absolute		kPaA
kilopascal gage (gauge)		kPag
kilovolt		kV
kilovolt ampere		kVA
kilowatt		kW
kilowatt hour		kWh
latitude	lat	
less than	<	
liter (litre)	L	L
liter per second		L/s
low pressure	lp	
low voltage	lv	
lubricate	lub	
lumen		lm
lumens per watt		lm/W
manager	mgr	
manufacturer	mfr	
mass	(spell out)	
material	matl	
maximum	max	
mechanical	mech	
mega (million)		M
megacycle	(spell out)	
megajoule		MJ
megapascal		MPa
megavolt		MV
megawatt		MW
megohm		MΩ
meter (metre)		m

	Abbreviation	Unit Symbol
meters per second		m/s
microampere		µA
milli (thousand)		m
milliampere		mA
milligram		mg
milliliter		mL
millimeter		mm
millimeter square		mm^2
million (mega)		M
millisecond		ms
millivolt		mV
minimum	min	
minute	min	
minute (angular measurement)		'
molecular weight	mol wt	
month	mo (or spell out)	
more than	>	
negative	neg	
Newton		N
nominal	nom	
not applicable	N/A	
number	no.[1]	
ohm	(spell out)	Ω
original	orig	
outside diameter	OD	
page	p	
paragraph	para	
parts per million	ppm	
pascal		Pa
pascal absolute		PaA
pascal gage (gauge)		Pag
per annum	pa	

(continued)

[1]no. with period is used in equipment data sheets in this book to differentiate it from the word "no", e.g., no/yes

	Abbreviation	Unit Symbol
percent	pct	%
phase	(spell out)	∅
power factor	pf	
pressure	press	
pressure drop	P.D.	
quantity	qty	
radian		rad
radius	r	
reference	ref	
relative humidity	rh	
revision	rev	
revolutions	rev(s)	
revolutions per minute		rpm or rev/min
revolutions per second		rev/s
second	sec	s
second (angular measurement)		″
section	(spell out)	
specific gravity	sp gr	
specific heat	sp ht	
specification	spec	
square	sq	
square centimeter		cm^2
square kilometer		km^2
square meter		m^2
square millimeter		mm^2
standard	std	
subject	subj	
table (tabulation)	(spell out)	
temperature	temp	
thousand		k
tonne		t
tonne per year		t/yr

	Abbreviation	Unit Symbol
tonne per day		t/day
tonne per hour		t/h
velocity	vel	
vertical	vert	
volt		V
volt ampere		VA
voltage drop		mV/A.m
volume	vol	
watt		W
watt hour		Wh
week	(spell out)	
with	w/...	
without	w/o	
year	yr	

INDEX

About the Author

Robert Leeser, until his retirement from full-time practice in 1989, has had 36 years of experience in engineering and contract supervision including the design, procurement, construction, and plant start-up of a wide range of projects for power, processing, and mining facilities.

He received a B.E. degree in mechanical engineering from the University of New South Wales, completed postgraduate studies in England, and worked as a project and commissioning engineer in steam and power plants before taking up engineering positions in the petrochemical industry. He then joined the Bechtel Corporation in Melbourne and later became Chief Mechanical Engineer with the Ralph M. Parsons Company, Australia. Prior to his retirement, he spent nine years in the coal mining industry involved in preparing procurement documentation and the selection of major, technologically advanced equipment for large, high-production opencut and underground mining, and material handling contracts.

Robert Leeser has authored several technical papers and is a member of the Institution of Engineers, Australia, a fellow of the Australian Institute of Energy, and a member of the American Society of Mechanical Engineers.